Berliner ökophysiologische

und phytomedizinische Schriften

Hrsg. von Christian Ulrichs und Carmen Büttner

Lebenswissenschaftliche Fakultät,

Humboldt-Universität zu Berlin

Band 51

Fresh-cut apples: Aspects of respiration, sanitation and storage conditions affecting quality and volatile synthesis

DISSERTATION

zur Erlangung des akademischen Grades
Doctor rerum agrarium
(Dr. rer. agr.)

eingereicht an der

Lebenswissenschaftlichen Fakultät der Humboldt-Universität zu Berlin

von

Herr Guido Rux

Präsidentin der Humboldt-Universität zu Berlin
Prof. Dr.-Ing. Dr. Sabine Kunst

Dekan der Lebenswissenschaftliche Fakultät
Prof. Dr. Dr. Christian Ulrichs

Gutachter/in:
1. Prof. Dr. Dr. Christian Ulrichs
2. Prof. Dr.-Ing. habil. Robert Kabbert

Datum der Einreichung: 17.07.2020
Datum der Promotion: 23.03.2021

Bibliografische Information der Deutschen Nationalbibliothek
Die Deutsche Nationalbibliothek verzeichnet diese Publikation in der Deutschen Nationalbibliografie; detaillierte bibliographische Daten sind im Internet über http://dnb.d-nb.de abrufbar.
1. Aufl. - Göttingen: Cuvillier, 2022
Zugl.: Berlin, Univ., Diss., 2021

Nonnenstieg 8, 37075 Göttingen
Telefon: 0551-54724-0
Telefax: 0551-54724-21
www.cuvillier.de

1. Auflage, 2022
Gedruckt auf umweltfreundlichem, säurefreiem Papier aus nachhaltiger Forstwirtschaft.

ISBN 978-3-7369-7564-4
eISBN 978-3-7369-6564-5

Summary

Due to the inevitable wounding and tissue destructions during minimal processing, fresh-cut fruits became highly perishable and show a short shelf life. In fresh-cut apples, which are often a major component of fruit salads, microbial loads, rapid tissue browning and loss of texture and flavour limit the customers' acceptability. Postharvest storage of fresh-cuts in sugar-syrup or suitable fruit juice solutions are common practices to extend the shelf life. However, knowledge on the effects of these applications on the quality attributes of fresh-cut fruits is sketchy. Low O_2 atmospheres, as they are found after immersion, may negatively affect flavour and aroma of fresh produce. Because microbial spoilage is one of the main factors limiting the shelf life of fresh-cut apples, adequate and gentle sanitation is essential. Short-term hot-water treatment (sHWT) was described as an effective method to reduce microbial contaminations. For fruit salad production, however, sHWT needs to be optimised to prevent fruit injury and reduction of produce quality, and to guarantee the commercially required shelf life of 10 d.

Objective of this thesis was the evaluation of the effects of complete immersion of fresh-cut 'Elstar' apple slices in sugar solution or orange juice on quality attributes (colour, texture, acidity and sugar content) and physiology (respiratory behaviour). Furthermore, this thesis analysed the potential synergistic interactive effects caused by sugar syrup immersion of apple slices and the O_2 availability on VOCs emissions. In addition, the potential implications of sHWT in the temperature range of 55–65 °C on quality attributes, fruit aroma and product physiology of immersed fresh-cut 'Braeburn' apple slices were investigated.

Results showed that, irrespective of the medium used, immersion of fruit slices created "modified atmospheres" inside the tissues, which directly retarded physiological processes. The immersion can effectively prevent tissue browning of 'Elstar' and 'Braeburn' apple slices. However, the use of orange juice as a preservation medium appears unsuitable for the safe storage of apple slices over several days. For the storage of apple slices in sugar syrups, sugar contents of 13.4 – 20 % yielded optimal results. A sugar concentration of 20 % guaranteed a shelf life of at least 10 d if fruit slices are stored at 4 °C. The immersion probably led to a pronounced loss of aroma but also prevented off-odour. An inhibition of the alcohol acyltransferase enzyme was partially observed, which is necessary for ester formation. This effect was independent of the O_2 availability.

However, immersion also enhanced microbial growth, where yeasts were identified as a critical parameter. It was shown that the microbial quality of the syrup also plays a major role, because microorganisms can contaminate the sugar solutions, adherent to the fruit slices. This highlighted the great importance of the initial hygienic state of the apples to be processed. Pre-processing sHWT at both 55 °C and 65 °C significantly reduced the initial microbial load of apples, comparable to post-processing acid treatment, but could not extend the shelf life of apple slices. Although sHWT induced marginal heat stress, this application doesn't negatively affect important quality parameters at both temperatures. In fact, sHWT might replace the post-processing acid treatment for fresh-cut products, because a combination of both technics did not enhance the reduction of microbes or further improve quality maintenance. The results support the conclusion that the direct heat inactivation of microorganisms is not the main mechanism of HWT sanitation and that the temporary melting of the epidermal wax layer may play an important role.

Zusammenfassung

Frisch geschnittene Früchte sind aufgrund der unvermeidlichen Verletzung des Gewebes bei der Verarbeitung leicht verderblich und nur sehr kurze Zeit haltbar. Bei frisch geschnittenen Äpfeln, die oft ein Hauptbestandteil von Obstsalaten sind, beschleunigt die Verarbeitung das mikrobielle Wachstum, die Gewebebräunung und den Verlust von Textur und Geschmack was alles zusammen zu einer abnehmenden Akzeptanz beim Kunden führt. Um die Haltbarkeit zu verlängern ist es gängige Praxis die Fruchtstücke in Zuckersirup oder geeigneten Fruchtsaftlösungen zu lagern. Die Kenntnis über mögliche Einflüsse dieser Lagermethode auf charakteristische Qualitätsmerkmale frisch geschnittener Früchte ist jedoch nur sehr lückenhaft. Die geringe O_2-Verfügbarkeit nach dem Eintauchen kann zudem den Geschmack und das Aroma dieser Produkte negativ beeinflussen. Da der mikrobielle Verderb einer der Hauptfaktoren bei der Haltbarkeitsbegrenzung von frisch geschnittenen Äpfeln ist, ist eine angemessene, aber produktschonende Hygienisierung unerlässlich. Seit vielen Jahren werden Heißwasserbehandlungen (HWT) als eine wirksame Methode zur Reduzierung mikrobieller Kontaminationen beschrieben. Für die Obstsalatproduktion kann jedoch nur eine optimierte kurzeitige Heißwasserbehandlung (sHWT) genutzt werden, um die von Produzenten geforderte Haltbarkeit von 10 Tagen zu garantieren ohne Qualitätsverluste bei geschnittenen Früchten zu riskieren.

Zielsetzung dieser Arbeit war es deshalb, die potentiellen Einflüsse der Lagerung von frisch geschnittenen 'Elstar'-Apfelscheiben in Zuckerlösung oder Orangensaft auf deren Qualitätsmerkmale (Farbe, Textur, Säure- und Zuckergehalt) und Physiologie (Respiration) umfassend zu analysieren und zu bewerten. Darüber hinaus wurden mögliche interaktive Effekte der Lagerung von Apfelscheiben in Zuckersirup und der damit verbundenen reduzierten O_2-Verfügbarkeit auf die Emissionen geruchsbildender Verbindungen untersucht. Weiterhin wurden die möglichen Auswirkungen einer sHWT im Temperaturbereich von 55 – 65 °C auf die Qualität, das Fruchtaroma und die Physiologie von frisch geschnittenen und in Zuckerlösung gelagerten 'Braeburn'-Apfelscheiben bewertet.

Die Ergebnisse zeigten, dass sich, unabhängig vom verwendeten Medium, durch das Eintauchen der Apfelscheiben innerhalb des Gewebes eine "modifizierte Atmosphäre" bildete, welche die physiologischen Prozesse der Apfelscheiben verlangsamte. Das Eintauchen konnte auch, wiederum unabhängig vom verwendeten Medium, die Gewebebräunung von 'Elstar'- und 'Braeburn'-Apfelscheiben wirksam verhindern. Allerdings scheint die Verwendung von Orangensaft als

Konservierungsmittel für eine sichere Lagerung von Apfelscheiben über mehrere Tage ungeeignet zu sein, da dessen mikrobieller Verderb schnell voranschreitet. Bei der Lagerung von Apfelscheiben in Zuckersirupen führten Zuckergehalte von 13 – 20 % zu den besten Ergebnissen. Eine Zuckerkonzentration von 20 % garantierte eine Haltbarkeit von mindestens 10 Tagen, wenn die Apfelscheiben bei 4 °C gelagert wurden. Das Eintauchen führte zwar zu einer ausgeprägten Reduzierung der Emissionen wichtiger Aromastoffkomponenten, verhinderte jedoch auch die Entwicklung von Fremdgerüchen. Dabei wurde teilweise eine von der O_2-Verfügbarkeit unabhängige Aktivitätshemmung des für die Ester-Bildung notwendigen Enzyms Alkohol acyltransferase beobachtet.

Die Lagerung in Zuckersirup verstärkte das mikrobielle Wachstum, wobei Hefen der kritischste Parameter waren. Es zeigte sich, dass auch die mikrobielle Qualität des Sirups eine große Rolle spielt, da die Zuckerlösung durch Mikroorganismen, welche an den Fruchtscheiben anhaften, kontaminiert werden kann. Dies unterstreicht auch die große Bedeutung des mikrobiellen Ausgangszustandes der zu verarbeitenden Äpfel. Die sHWT reduzierte sowohl bei 55 °C als auch bei 65 °C den mikrobiellen Besatz der Äpfel deutlich. Diese Reduktion war vergleichbar mit der praxisnahen Anwendung von Ascorbin- und Zitronensäurelösungen nach dem Schneiden. Die sHWT konnte die Haltbarkeit der Apfelscheiben jedoch insgesamt nicht weiter verlängern. Obwohl die sHWT einen geringfügigen Hitzestress induzierte, zeigte diese Anwendung bei beiden getesteten Temperaturen keinen negativen Einfluss auf wichtige Qualitätsparameter. Tatsächlich könnte die sHWT die zusätzliche Nutzung von Säurelösungen bei frisch geschnittenen Produkten ersetzen, da eine Kombination beider Techniken weder eine Reduzierung von Mikroorganismen noch eine bessere Erhaltung der Qualität bewirkte. Die erzielten Ergebnisse stützen auch die Schlussfolgerung, dass die direkte Temperatureinwirkung der sHWT nicht der Hauptmechanismus der Inaktivierung von Mikroorganismen ist. Hierbei spielt wahrscheinlich das vorübergehende Schmelzen der epidermalen Oberflächenwachsschicht der Äpfel eine wichtige Rolle.

Contents

List of publications

This thesis contained the following articles, referred to Roman numerals in the text:

I Rux, G., Caleb, O.J., Fröhling, A., Herppich, W.B., & Mahajan, P.V., (2017). Respiration and storage quality of fresh-cut apple slices immersed in sugar syrup and orange juice. *Food Bioprocess Technol.*, 10, 2081-2091. https://doi.org/10.1007/s11947-017-1980-6

II Rux, G., Bohne, K., Ulrichs, C., Huyskens-Keil, S., Hassenberg, K., & Herppich, W.B., (submitted 2020). Effects of modified atmosphere and sugar immersion on physiology and quality of fresh-cut 'Braeburn' apples. *Food Packag. Shelf Life.*

III Rux, G., Efe, E., Ulrichs, C., Huyskens-Keil, S., Hassenberg, K., & Herppich, W.B., (2019). Effects of pre-processing short-term hot-water treatments on quality and shelf life of fresh-cut apple slices. *Foods*, 8 (12), 653. https://doi.org/10.3390/foods8120653.

IV Rux, G., Efe, E., Ulrichs, C., Huyskens-Keil, S., Hassenberg, K., & Herppich, W.B., (2020). Effects of pre-processing hot-water treatment on aroma relevant VOCs of fresh-cut apple slices stored in sugar syrup. *Foods*, 9 (1), 78. https://doi.org/10.3390/foods9010078

The contribution of Guido Rux to the articles included in this thesis was as follows:

Developed the concept and the study design in collaboration with the co-authors (I-IV). Planned the experiments together with the co-authors (I-IV). Performed all of the experimental work (I) or performed major parts of the experimental work and supervised the students in performing some parts of the experimental work (II-IV). Data collection and analyses/evaluation (I-IV). Mainly wrote the manuscript (I-IV). All co-authors contributed with invaluable ideas for the data analyses, provided valuable support to results evaluation and gave helpful suggestions for the manuscript.

Funding: The current research was funded by the German Federation of Industrial Research Associations—"Otto von Guericke" e.V. (AiF)/German Federal Ministry for Economic Affairs and Energy (BMWi) (NO:F4001902MD5), within the research program Zentrales Innovationsprogramm Mittelstand (ZIM). Financial support provided by the Georg Forster Postdoctoral Research Fellowship (HERMES) programme from the Alexander von Humboldt Foundation (Ref. 3.4–ZAF–1160635-GFHERMES-P) is appreciated.

Abbreviations

3-2EI	trans-2-enal isomerase
AAT	alcohol acyltransferase
acyl-CoA	acyl-coenzyme A
acyl-CoAs	acyl-coenzyme A (shorter chain)
ADH	alcohol dehydrogenase
AER	alkenal oxidoreductase
AOR	NADPH-dependent alkenal/one oxidoreductase
ATP	adenosine triphosphate
BCAT	branched-chain amino transferase
BCKDH	branched-chain A-keto acid dehydrogenase
CA	controlled atmosphere
FAD	flavin adenine dinucleotide
GC	gas spectrometer
GDH	glutamate dehydrogenase
HDH	2-hydroxy-3-methyl-pentanoic acid dehydrogenase
HPL	hydroperoxide lyase
HWT	hot-water treatment
Ile	isoleucine
Leu	leucine
LOL	low oxygen level
LOX	lipoxygenase
MA(P)	modified atmospheres (packaging)
MS	mass spectrometer
NAD(H\H+H+)	nicotinamide adenine dinucleotide
NADPH	nicotinamide adenine dinucleotide phosphate
PAT	phosphate acetyltransferase
PCA	principal components analysis
PDC	pyruvate decarboxylase
PDH	pyruvate dehydrogenase

PPO	polyphenol oxidase
RI	retention indice
RR	respiration rate
sHWT	short-term hot-water treatment
SPME	solid phase microextraction
SSC	soluble solid content
TA	(total) titratable acidity
TCA	tri-carboxylic acid
ULO	ultra-low oxygen
Val	valine
VOC	volatile organic compound

List of tables

List of figures

1 Introduction

The fresh-cut fruits sector has shown consistent growth in the last few years, due to changes in consumers' lifestyle and the growing demand for convenient, nutritional and safe "ready-to-eat" fruits (Rabobank, 2010; Baselice et al., 2015). The strongest incentive to buy fresh-cut products is convenience (Ragaert et al., 2004). Restaurants, fast-food outlets and institutional food-service operators seeking to reduce labour costs by buying minimally processed, ready-to-use or ready-to-eat fresh-cut fruits further boost this demand. According to Beaulieu (2010), the higher prices of fresh-cut produce will be accepted by customers, if the quality is equal to the uncut product. However, due to the inevitable wounding and tissue destructions during minimal processing, fresh-cut fruits become highly perishable and show a short shelf life. In fresh-cut apples, which are often a major component of fruit salads, microbial loads, rapid oxidative tissue browning and loss of texture and flavour are important factors limiting the customers' acceptability (Dixon and Hewett, 2000; Toivonen and DeEll, 2002).

To optimize the shelf life of fresh-cut fruit, various postharvest treatments have been investigated or applied over the past years to minimize the adverse impact of minimal processing. These include the use of modified atmosphere (MA) and humidity packaging (Rux et al., 2015; Caleb et al., 2016), edible coatings (Rojas-Graü et al., 2007; Rojas-Graü et al., 2008; Salvia-Trujillo et al., 2015), thermal (e.g. Koukounaras et al., 2008) and plasma (e.g. Tappi et al., 2014) treatment, UV-C combined with MA-packaging (López-Rubira et al., 2005) or the application of organic acid solutions and sugar (Verlinden and Garcia, 2004; Nishikawa et al., 2005). In particular, chemicals (Buta et al., 1999) and/or MA during storage (Gorny et al., 1998) are used to control oxidative browning and to delay undesirable biological and biochemical changes (Beaulieu, 2010). Postharvest sugar-syrup application or storing fresh-cuts in suitable fruit juice solutions as an additive are also general practices.

The immersion of fresh-cut fruits in sugar syrup is a common practice especially used for bulk purchasers (Alzamora et al., 1993) to extend the common maximum shelf life for fresh-cut fruit salads of 5 d (Seide et al., 2017) to up to 10 d. According to practical experience, storage of fruits in sugar syrup (ca. 20 %) extend product shelf life by preventing enzymatic and oxidative browning and transpiration and slows down respiration, ethylene metabolism and other physiological processes. Despite reports on the effects of storage in sugar syrup on ethylene metabolism (Nishikawa et al., 2005) or chlorophyll degradation of fresh-cut fruit (Coupe et al., 2003), the knowledge on the impact of sugar solution on the physiological properties of immersed fresh-cuts is still limited. However, information on the physiological responses and changes in quality of fresh-cuts immersed in sugar solution or fruit juice is lacking. The current industrial practices are based on "trial-and-error approach" without an adequate evaluation of the sugar content or juice concentrations. This may result in non-optimal shelf life and quality, which cumulates in economic and postharvest losses. For instance, low O_2 atmospheres, as they are found after immersion, can negatively affect flavour and aroma (El Hadi et al., 2013). The aroma is comprised by a high number of volatile organic compounds (VOCs) (DeRovira, 1996; Baldwin et al., 2007) and is important for the consumers' sensorial acceptance (Beaulieu, 2010).

According to the fresh-cut producer, microbial spoilage is the main factor limiting the shelf life of apple slices immersed in sugar syrup. Consequently, careful removal of microorganisms, adherent to fruit skin (Pietrysiak and Ganjyal, 2018) is essential to avoid microbiological contamination and cross-contaminations (Kabelitz and Hassenberg, 2018). In practice, apple slices are often dipped in a mixture of citric and ascorbic acid solutions before immersion to prevent browning and for sanitation purposes. To prevent the consumption of these chemicals especially in organic production, gentle physical sanitation methods are demanded.

Hot-water treatment (HWT) in the temperature range of 40 – 80 °C may effectively reduce microbial contaminations, is relatively inexpensive and is easy to use (Lurie, 1998). In addition, HWT maintains storage quality of fruits (Shao et al., 2007; Spadoni et al., 2015; Kabelitz and Hassenberg, 2018). Being chemical-free, particularly short-term (15 to 60 s) hot-water treatment (sHWT) is suitable for organic production (Fallik, 2004; Maxin et al., 2014). Consequently, the application of short-term hot-water treatments (sHWT) of apples before processing could represent an effective but gentle sanitation technique to additional prolong their shelf life. Besides earlier studies on the impacts of HWT on structure and function of fruit epidermal tissue (Roy et al., 1994; Lurie et al., 1996), only recently, the effects of sHWT on surface tissue, heat transfer dynamics and suitability for pre-processing of intact apples for fresh-cut salads were investigated in detail (Kabelitz and Hassenberg, 2018; Kabelitz et al., 2019). For fruit salad production, however, sHWT needs to be optimised to prevent fruit injury and reduction of produce quality, and to guarantee the commercially required shelf life of 10 d.

This study aims to evaluate systematically the impacts of postharvest storage in sugar-syrup and sHWT on important quality parameters and quality relevant biosynthetic pathways, and microbial loads of fresh-cut apples. However, both methods are not investigated adequately and the knowledge on their functional mechanisms, their effectiveness and their potential effects on fresh-cut fruits' quality and shelf life is sketchy. A special emphasis is put on the potential synergistic interactive effects caused by sugar syrup immersion of apple slices and restricted O_2 availability on the synthesis and the emission of VOCs.

The objective of the first scientific part of this thesis, presented in chapter 3, was a comprehensive scientific evaluation of postharvest storage in sugar-syrup. In this chapter, the impact of the complete immersion of fresh-cut apples in sugar syrup and orange juice solutions on the respiratory behaviour and on commonly used quality attributes (colour, texture, acidity and sugar content) was investigated. Special focus was laid on the optimisation of the respective concentrations to increase quality or extend the shelf life. Next to the quality factors, chapter 4 focuses on the potential synergistic interactive effects caused by the sugar syrup immersion of apple slices and the O_2 availability on development and emission of VOCs. Chapter 5 and 6 are focusing on the potential implications of sHWT on the physiological quality parameters (5) and fruit aroma development (6) of immersed fresh-cut 'Braeburn' apple slices. The results will allow assessing whether sHWT could potentially supplement or replace post-processing chemical treatments. Additional, it will enable comprehension of the respective quality-related physiological processes to effectively select the optimal process conditions in processing of ecologically produced fresh-cut fruit salads. In the following chapter 7, important aspects of fresh-cut apple metabolism and investigated sanitation methods are summarized. In addition, the results of the different investigations are comprehensively discussed.

2 Scientific background

The fresh-cut produce is different from intact fruit in terms of physiology and postharvest handling requirements. Processing of fresh-cut fruit salads involves sorting, cleaning, washing, peeling, deseeding/coring and cutting. The initial responses of fresh-cut produces to wounding include a strong temporary increase in both respiration and ethylene production (Mahajan et al., 2014) mainly due to the initial stress response (Finnegan et al., 2013). The inevitable wounding and tissue destructions during minimal processing render fresh-cut fruits highly perishable and reduce their shelf life. Peeling or slicing enlarges the surface and leads to the release of cell contents (Beaulieu, 2010). This facilitates the spread and growth of naturally occurring microorganisms, normally only located on the fruit skin (Nicoli et al., 1994). Mixing of enzymes and the cytoplasmic and nucleic substrates, normally sequestered in different cell compartments, also leads to enzymatic browning of cut surfaces. The release of secondary fermentative metabolites also may negatively affect fruit aroma (Artés et al., 2007). Furthermore, peeling or slicing intensifies water losses, thus, increasing the risk of wilting and dehydration (Toivonen and DeEll, 2002). This effect results from both artificially enhanced surface and reduced diffusion resistance due to the removal of the natural protecting epidermal layer (Beaudry, 2000; Watkins, 2000).

Additional to this, fresh produce remain physiologically active and, thus, mechanical injury of plant tissues, induces a catena of physiological responses (Watada and Qi, 1999; Beaulieu, 2010), including the increase in respiration and in the biosynthesis of ethylene and secondary metabolites, which finally results in the loss of quality and aroma (Toivonen and DeEll, 2002). Additional, the activation of enzymatic systems may accelerate cell membrane degradation and shorten the shelf life of fresh-cut produces (Toivonen and DeEll, 2002; Nicoli et al., 1994). In fresh-cut apples, which are often a major component of fruit salads, also cultivar-specific properties, stage of maturity at cutting, and storage atmosphere and temperature (Gorny et al., 1998) further affect the shelf life.

2.1 PPO-mediated apple tissue browning

One of the most important quality parameters for fresh-cut apples is the colour of the cut tissue, which essentially affects the appearance and thus has a strong impact on the consumer's buying decision (Toivonen and Brummel, 2008). Therefore, preservation of normal tissue colour and control of discolouration or surface browning is an important consideration during fresh-cut fruit processing and storage. Tissue browning due to enzymatic action mainly changes the lightness (L*) and the greenness-redness (a*) of the pulp. The enzymatic action is induced by tissue wounding through the slicing of apples and associated destruction of cell membranes inside plant tissues, which leads to the mixing of phenolic compounds with the endogenous polyphenol oxidase (PPO) (Gil et al., 1998; Toivonen, 2004). Therefore, enzymatic browning is mainly determined by the PPO activity and the phenolic compound concentrations (Martinez and Whitaker, 1995). Further factors are the pH, the temperature and the oxygen availability of the tissue (Martinez and Whitaker, 1995). The oxygen availability is considerably increased by cutting due to the reduced gas diffusion resistance (Gil et al., 1998).

According to Harel et al. (1964), catechol oxidases are the most common polyphenol oxidases in apple fruit. Generally, PPO catalyses the oxidation of phenolic compounds to the corresponding o-quinone (Gacche et al., 2006), for which oxygen is needed. According to Cortellino et al. (2015), PPO hydroxylates monophenols to colourless diphenols and oxidates diphenols to coloured quinones. The hydroxylation is relatively slow, whereas the oxidation is relatively rapid. The brown or black melanin pigments, which are associated with "browning" in plant tissues, are accumulating through subsequent reactions of the quinones (Toivonen and Brummel, 2008). In these processes, the specific structure of present polyphenolic substrates determines the formation and sequence of brown or black-coloured reaction products (Toivonen and Brummel, 2008).

In the past, several commercial and research strategies were described to reduce PPO-mediated discolouration (Gorny, 2001; Garcia and Barrett, 2002; Hodges and Toivonen, 2008). According to Gorny (2001), reduced oxygen and/or elevated carbon dioxide concentrations may prevent enzymatic browning of fresh-cut fruits and vegetables. Soliva-Fortuny et al. (2002) compared 'Golden Delicious' apple slices stored (60 d) in air or under 100 % N_2 and reported slightly lower L* values of the samples in MA storage. Prevention of browning the modified gas atmosphere was much more effective when combined with a preceding ascorbic acid dip. In this context, Cortellino et al.

(2015) also confirmed a colour preservative effect of MA only for the samples dipped in anti-browning solutions before storage.

The application of reducing agents such as ascorbic acid (and its derivatives) and citric acid can inhibit the browning effects of fresh-cut fruits (Nicoli et al., 1994; Soliva et al., 2001). Ascorbic acid is commercial often used as an anti-browning agent due to its versatile action (Luo and Barbosa-Canovas, 1996; Buta et al., 1999; Cocci et al., 2006; Tortoe et al., 2007). It chelates copper ions, reduces o-quinones to their original polyphenolic compounds and acts as a competitive PPO inhibitor (Lozano-de-Gonzales et al., 1993). For example, Gil et al. (1998) showed the effectiveness of 2 % ascorbic acid as a reducing agent, which prevented the decrease of total polyphenol content during storage of 'Fuji' apple slices.

According to Rojas- Graü et al. (2006), the maturity stage of the processed apples affected the effectiveness of ascorbic acid as a browning inhibitor. It was pointed out that ascorbic acid, as a reducing agent, is consumed during anti-browning reactions (Luo and Barbosa-Canovas, 1997). However, according to Chiabrando and Giacalone (2012), the effectiveness of ascorbic acid in the inhibition of browning is lower than that of citric acid.

Citric acid is also known to prevent browning and to maintain the quality of fresh-cut samples (Jiang et al., 2004; Queiroz et al., 2011). The inhibitory effect of this agent may be related to the phenolase Cu-chelating power (Tortoe et al., 2007). To get a more effective browning preservation of fresh-cut fruits, different substances can be combined (Ahvenainen, 2000). In this context, ascorbic acid (antioxidant agent) has been long applied in combination with citric acid (acidulant) to prevent enzymatic browning of sliced apples (Ponting, 1972; Sapers and Douglas, 1987; Pizzoccaro, 1993; Laurila et al., 1998; Soliva-Fortuny and Martín-Belloso, 2003a).

In this context, it has been shown that solutions of acidic pH or the addition of acidulates may effectively inhibit or control enzymatic browning (Nicolas et al., 1994; Whitaker, 1994; Chiabrando and Giacalone 2012). Biegańska-Marecik (2007) suggests using 0.8 % citric acid in the dipping solution in combination with sucrose, to inhibit effectively colour changes in apple slices.

The combination of anti-browning dippings and storage in modified atmospheres were investigated in numerous studies. Ascorbic acid/citric acid-treated apple slices retained colour attributes better when additional MA-stored at 4 °C for 8 d (Cocci et al., 2006) or 11 d (Cortellino et al., 2013) than when stored in air. Similar results were reported by Gunes et al. (2001; 0 – 1 % O_2), Soliva-Fortuny et al. (2002; 100 % N_2) and

Rocculi et al. (2004; 5 % O_2 + 5 % CO_2 + 90 % N_2). Furthermore, Soliva-Fortuny et al. (2001) described a decrease in the PPO activity at additionally enhanced CO_2 concentration (2.5 % O_2 + 7 % CO_2 + 90.5 % N_2).

2.2 Changes in the texture of apple slices

The term texture summarizes the structural and mechanical properties of a product, as well as its sensory properties, which can be perceived by hand and mouth (Bourne, 2002; Abbott and Harker, 2004; Barrett et al., 2010; Beaulieu, 2010). Softening is an undesirable consequence of cutting resulting from changes in the physical and the mechanical properties, and in the chemical structure of the tissue. On the other hand, firmness is a sensorial property, which generally describes the impression and feeling of texture in sensory testing (Bourne, 2002; ASABE, 2008), but often misinterpreted as stiffness or tissue strength (Vincent, 1994). The latter is determined by cell size, biochemical and biophysical cell wall properties, cell-to-cell adhesion and tissue turgor (Toivonen and Brummel, 2008). The loss of tissue strength is associated with changes of the cell wall pectins to water-soluble pectins by pectinolytic hydrolysis via polygalacturonases and/or pectin methyl esterases, decreasing cellulose content crystallinity of cellulose fibrils, thinning of cell walls, diffusion of sugar to the intercellular spaces and ion movement from the cell wall (Toivonen and Brummel, 2008; Qi et al., 2011). This may be due to highly controlled developmental processes during maturation or due to degradation reactions e.g. following mechanical stresses of plant tissue (Varoquaux et al., 1990). The changes in structure and biochemistry of pectins play a key role in fruit softening (Fischer and Bennett, 1991). Furthermore, wound-induced ethylene synthesis may also promote development and also induce senescence processes resulting in the alterations of cell wall metabolism especially for climacteric fruits (Gorny et al., 2002). In addition, water losses intact but particularly in fresh-cut apples due to the removed cuticle and sub-epidermal layers may lead to a decrease in turgor and results in loss of stiffness (Toivonen and Brummel, 2008).

According to Varoquaux and Wiley (1997), MA conditions reduce tissue destruction. Mainly dependent on the availability of O_2, this inhibits the loss of compartmentation within cells and diminishes the interaction between enzymes and their substrates (Soliva-Fortuny et al., 2002, 2003b and 2005). For example, Cortellino et al. (2015) showed a better preservation of the firmness of 'Golden Delicious' apple slices during storage at different MA conditions (1 % O_2 + 99 % N_2; 5 % O_2 + 5 % CO_2 + 90 % N_2) than during air-storage.

Results of Cocci et al. (2006) indicated a very minor lower firmness of 'Golden Delicious' apple slices stored in air compared to MA-stored samples (1 % O_2 + 2 % CO_2). Storage of apple slices at low O_2 concentrations may prevent softening for up to 3 weeks (Soliva-Fortuny et al., 2005). In this context, Soliva-Fortuny et al. (2003b) evaluated the microstructural modification of fresh-cut apples stored under different atmospheric conditions using Cryo-Scanning Electron Microscopy. They detected progressed cell deterioration for samples stored under MA conditions (2.5 % O_2 + 7 % CO_2) and indicated by a great number of exudates formed as droplet-shape on the external surface of the cell walls. Samples stored under LOL conditions (100 % N_2), however, showed additionally smoother and more integer cell walls, ranked between fresh apples and MA samples. The authors attributed this to the maintenance of the aerobic metabolism, which may finally degrade the fruit tissue.

However, studies also suggested softening of apple slice tissues due to the structural breakdown by the application of acid solutions (Ponting et al., 1971; Gil et al., 1998; Rojas-Graü et al., 2007; Cortellino et al., 2015). In this context, experiments of Cocci et al. (2006) showed that firmness of dipped apple slices decreased from 8.5 to 4.5 N within just one day of cold storage. Furthermore, the degree of softening was greatly affected by the mature stage of processed apples (Ponting et al., 1972; Rojas-Graü et al.; 2007). Apple slices dipped in ascorbic acid-solution rapidly lost firmness when they had been processed from ripe fruit, whereas this effect was less pronounced for apples in an advanced ripeness state and not observed for mature-green fruit (Rojas-Graü et al., 2007). This effect is related to the increased ethylene synthesis during ripening of these climacteric fruit, which is associated with the accelerated senescence leading to alterations in the cell wall metabolism (Gorny et al., 2002). In this context, low O_2 concentrations successfully preserved the firmness of apple slices by limiting the ethylene production (Cortellino et al., 2015). This positive effect, however, was almost completely neutralized by the acid-dipping, which caused structural breakdown. Biegańska-Marecik (2007) also found of the tissue of apple slices softened after the addition of 0.8 % citric acid to the dipping solution.

2.3 Quality relevant apple constituents

Flavour and nutritional value of fresh-cut apples are important quality factors for consumers (Beaulieu, 2010) and is mainly related to their natural ingredients. The sensory attribute flavour generally comprises aspects of both aroma and taste. According to Beaulieu (2010), flavour represents a very complex “trait” that is difficult to analyse because it is created by a large variety of very different compounds such as sugars, salts, acids, alkaloids, flavonoids and VOCs (DeRovira, 1996; Baldwin et al., 2007). Furthermore, sensory attributes such as flavour or the nutritional value of stored produce may be lost prior the deterioration of visible appearance (Beaulieu, 2010). As an example, the senescence of fresh-cut fruit may directly result in adverse changes of flavour that are concomitantly induced by distinct catabolic and/or metabolic mechanisms; it may also lead to changes in diffusional properties of tissues (Beaulieu, 2006, Forney, 2008). Therefore, the monitoring of flavour and nutritional relevant ingredients is general practice in fruit production to evaluate growing, postharvest treatments or storage conditions.

Apples consist of approx.. 850 g kg^{-1} water, as well as carbohydrates and fibres. In addition, apples contain numerous important micronutrients such as organic acids, vitamins, minerals (potassium, sodium, magnesium, calcium and iron), trace elements and secondary plant substances (e.g. phenols and carotenoids), 75% of which are found in or directly under the apple skin (Buchter-Weisbrodt, 1998). Most of these micronutrients have a positive effect on health (Hecke et al., 2006). As the flavour of an apple is strongly determined by the ratio of sugar to acidity (Hecke et al., 2006), both are relevant parameters in fruit industry. However, sour tasting cultivars are not necessarily low in sugar. This was shown, for example, by the regional ’Steirische Schafnase’ apple, which was classified as sour by a taste-test only because the extremely high sugar concentration was masked by the high acid content (Hecke et al., 2006). The specific contents of sugars and acids are highly cultivar-specific, but can vary according to weather and soil (Hecke et al., 2006). The main controllable factor influencing flavour is the maturity stage at harvest (Kader, 2008), which also essentially determines the quality and shelf life of fresh-cut fruits. Thus, harvesting immature or overripe fruits may result in poor flavour (Kader, 2002). The contents of sugar and organic acids directly depend on the maturity stage. During ripening, sugar contents increase, whereas acidity decreases, while the synthesis of flavour compounds occurs relatively late in fruit maturation (Garcia and Barrett, 2005).

Apples provide carbohydrates in the form of easily digestible sugars with contents between 115 and 200 g kg^{-1} (Hecke et al., 2006), usually measured as total soluble solid content (SSC). Fructose and sucrose are the major sugars, followed by lower proportion of glucose (5 – 20 g kg^{-1}), which all show a constant ratio to each other in the fruit (Hecke et al., 2006). During postharvest storage, the disaccharide sucrose will be hydrolysed to glucose and fructose (Chardonnet et al., 2003). However, the total sugar content of processed apples is not or just marginally affected by refrigeration, modified atmospheres or anti-browning treatment (Rocha et al., 1998; Buta et al., 1999; Bett et al., 2001; Rocculi et al., 2004). Sucrose pre-treatments of fresh-cut 'Jonagold' apple slices before vacuum packing were shown to increase the soluble solid contents depending on the sugar concentrations used (Biegańska-Marecik and Czapski, 2007).

Besides their importance for flavour, fruit acids are also valuable in preventing human illness and slight metabolic disorders (Hecke et al., 2006). Organic acids or total titratable acids (TA) in apples are mainly composed of malic acid, citric acid, oxalic acid and salicylic acid (Buchter-Weisbrodt, 1998). The contents of these compounds range between 6 and 14 g kg^{-1}, where malic acid represents the main component in fruit of almost all cultivars (Hecke et al., 2006). According to Hecke et al. (2006), the concentration of citric acid is cultivar-specific. These authors also found very low concentrations of fumaric acid, between 0.5 – 1.9 mg kg^{-1}, and shikimic acid below 0.05 g kg^{-1} in all cultivars. In storage of fresh-cut apples, organic acids declined drastically during storage at both 5 and 10 °C (Buta et al., 1999). Lamikanra et al. (2000) showed that the reduction in organic acids contents is highly influenced by the storage temperature and acid degradation may be prevented by refrigeration in cantaloupe melon. Cocci et al. (2006) reported no major change in TA content of fresh-cut 'Golden Delicious' stored under modified atmosphere (1 % O_2 + 2 % CO_2) at 4 °C for 8 d. Similar results were obtained by other studies (Soliva-Fortuny and Martín-Belloso, 2003a; Rocculi et al., 2004; Cliff et al., 2010).

While modified atmosphere conditions probably do not affect minimally processed apples, the effect of antioxidant treatment is diverse. Here, Cocci et al. (2006) reported an increase of TA in apple slices after dipping them for 3 min in antioxidant solution (1 % ascorbic acid plus 1 % citric acid) due to the uptake of the anti-browning solution, and slower and less intense increase in ripening index (SSC/TA). But Rocha et al. (1998) reported a decline of TA in fresh-cut apples after chemical treatments with calcium chloride, ascorbic acid and citric acid, due to the quick consumption of organic acids by increased respiration after peeling and cutting, as well as by microorganisms (Soliva-Fortuny et al., 2005).

Above all, apples are an important source of vitamin C (L-ascorbate). The ascorbate content depends on the apple cultivar, e.g. specified for 'Braeburn' and 'Elstar' with 240-350 mg kg^{-1} and 80-150 mg kg^{-1}, respectively (Buchter-Weisbrodt, 1998). Vitamin C is involved in many aspects of cellular metabolism (plant growth, development, hormone function, transcription); it is also the most important antioxidant in apple (Eberhardt et al., 2000). Therefore, vitamin C has benefits for the consumer, e.g. scavenge of free radicals in human cells (Schirrmacher and Schempp, 2003) and the potential to improve the storage properties (Eberhardt et al., 2000). Being water soluble, vitamin C degrades relatively quickly and represents one of the most sensitive vitamin; thus, it can be used as an index of nutrient content or its degradation (Garcia and Barrett, 2005). This is relevant, because the nutritional value is a quality factor that is difficult to evaluate and many factors contribute to the nutrient content of a fruit, including genetics, growing conditions, maturity at harvest, production practices and postharvest handling conditions (Garcia and Barrett, 2005). In apple processing, washing probably does not affect vitamin C (Ahvenainen, 1996), whereas cutting leads to rapid losses of certain vitamins due to increased respiration and senescence (Garcia and Barrett, 2005). The further degradation of vitamin C during storage is enhanced at increased temperatures (Garcia and Barrett, 2005). Storage under controlled atmosphere conditions also leads to losses of vitamin C, e.g. as shown for whole pears (Garcia and Barrett, 2005). This effect may be attributed to the high CO_2 concentrations, which stimulate the oxidation of ascorbic acid and/or the inhibition of their reduction from dehydroascorbic acid (Soliva-Fortuny and Martín-Belloso, 2003a).

2.4 Aroma volatiles (VOCs)

The aroma plays a highly important role in determining the perception and acceptance of products by the consumers (El Hadi et al., 2013). The characteristic aroma of fruits is mainly created by the emission of volatile organic compounds (VOCs) (Fellman et al., 2000; El Hadi et al., 2013). In apples, volatile aroma compounds represent a key quality attribute (Espino-Díaz et al., 2016) and gained increasing attention in recent years (El Hadi et al., 2013).

2.4.1 Aroma volatile composition, biosynthesis and characteristics and synthesis

In most fruits, VOCs are produced in significant numbers and form complex volatile profiles depending on the cultivar and stage of fruit ripeness (Berger et al., 1986; Bruckner, 2008; El Hadi et al., 2013). This also applies to apples (Dixon and Hewett, 2000; Fellman et al., 2000), for which more than 300 volatiles were identified (Nijssen et al., 2011).

This includes diverse groups of organic compounds such as alcohols, aldehydes, carboxylic esters, ketones and ethers (Dimick and Hoskin, 1983). Dixon and Hewett (2000) identified nearly 20 "character impact" VOCs as typical for aroma of apples (**Table 2.1**).

Table 2.1. Important apple (*Malus domestica*) volatile compounds, their aroma threshold ([a] Dixon and Hewett, 2000; [b] Burdock, 2016) values and sensory descriptions (adapted from Dixon and Hewett, 2000)

VOC	Threshold ($nL\ L^{-1}$)	Sensory description
Esters		
Butyl acetate	6.6[a]	red apple, Cox-like, harmonious, nail polish
Pentyl acetate	4.3-5[a]	banana like, apple, fruity, Gala
Hexyl acetate	2-115[a]	characteristic apple, Cox-like, sweet fruity, ripe, pear
2-methylbutyl acetate	5-11[a]	overall aroma, characteristic apple, solvent, banana like
Ethyl butanoate	1[a]	fruity, estery, harmonious, fruity
Ethyl 2-methylbutanoate	0.006-0.1[a]	Fruity, apple like, sweet strawberry
4-methoxyallylbenzene	-	spicy, aniseed
Methyl 2-methylbutanoate	0.92-4.4[b]	sweet fruity
Propyl 2-methylbutanoate	-	very sweet, strawberry
Butyl 2-methylbutanoate	17-87[b]	fruity, apple
Hexyl 2-methylbutanoate	22-430[b]	apple, grapefruit
Butyl hexanoate	700-10000[b]	green apple
Hexyl propanoate	8[b]	apple
Butyl butanoate	87-1000[b]	rotten apple, cheesy
Butyl propanoate	25-440[b]	fruity, apple
Hexyl butanoate	250[b]	apple
Hexyl hexanoate	6400[b]	apple
Aldehydes		
Acetaldehyde	15-120[a]	green/sharp
trans-2-hexenal	1-17[a]	green/sharp, harmonious, fruity
Hexanal	5[a]	green/sharp, earthy, grass like
Alcohols		
Butan-1-ol	500[a]	overall flavour, aroma, sweet aroma
Hexan-1-ol	150-500[a]	earthy, unpleasant
trans-2-hexenol	$1x10^{7}$ [b]	harmonious, fruity

The relevant sensory thresholds of some of these compounds are very low; therefore, they have a high impact on the characteristic aroma. Other VOCs, however, intensify the aroma or contribute to aroma quality (Dürr and Schobinger, 1981). Although the same VOCs are present in the volatile profiles of fruit of most apple cultivars, the sensorial experience of their specific aroma may pronouncedly vary due to the absence of characteristic key VOCs or to different proportions of these compounds (Poll, 1981; Cunningham et al., 1986; Paillard, 1990).

Different biochemical pathways are responsible for the synthesis of the distinct aroma compounds (El Hadi et al., 2013). In this context, mainly fatty and amino acids act as precursors and are predominantly converted into aldehydes, alcohols and esters (Espino-Díaz et al., 2016). Epidermal tissue produces higher amounts of VOCs than the residual fruit body (Rudell et al., 2002). This is mostly due to their higher metabolic activity or to the higher presence of fatty acids (Guadagni et al., 1971; Defilippi et al., 2005). The proportion of specific VOCs in the aroma profiles depends on substrate availability, and the substrate specificity and activity of relevant enzymes in the respective biosynthetic pathways (Rizzolo et al., 2006; El Hadi et al., 2013).

Biochemical reactions during climacteric ripening increase precursor availability and induced ethylene-related responses (Song and Bangerth, 1996; Fellman et al., 2000). The predominance of aldehydes in the VOC profiles of unripe apples changes as the contents of alcohols and esters considerably increase during progressing maturity. Especially the latter group of organic compounds dominate the final VOC profiles (Fellman et al., 2000). According to Paillard (1990), esters and alcohols can account 78 – 92 % and 6 – 16 % of typical profiles. Relevant esters are mostly synthesised from acetic, butanoic and hexanoic acids with ethyl, butyl and hexyl alcohols (Paillard 1990).

Both pre- and post-harvest conditions may affect the VOC profiles of apples (Berger et al., 1986; Bruckner, 2008). This includes the cultivar-specific genetic resources, growing conditions and cultural practices, the stage of maturity of the fruit, as well as postharvest handling and storage (El Hadi et al., 2013). During storage of intact or fresh-cut apples, temperature is an important factor affecting fruit aroma. Increased temperatures intensify the production of VOCs (Guadagni et al., 1971; Wills and McGlasson, 1971; Fallik et al., 1997), while specific VOCs may only be produced at certain temperatures (Guadagni et al., 1971; Dixon and Hewett, 2000). Additionally, VOCs production may temporarily be reduced by heat (e.g. $T > 46$ °C), due to inhibition or inactivation of relevant enzymes (Dixon and Hewett, 2000).

Furthermore, storage under modified atmosphere (MA) or controlled atmosphere (CA) conditions, used to maintain the quality of apples (Artes, 2006; Brecht, 2006; Erkan and Wang, 2006; Martin-Belloso et al., 2006; Yahia, 2006; Beaulieu, 2010), may decrease the quantity of produced aroma compounds (Patterson et al., 1974; Artes, 2006). Reducing the ethylene syntheses by low O_2 availability may retard the alteration of the VOC profile (Mattheis et al., 2005). Very low O_2 concentrations, however, may induce the anaerobic respiration and, thus, result in development of pronounced off-flavour (El Hadi et al., 2013). Additional, elevated CO_2 concentrations can inhibit the production of the majority of amino acid precursors and consequently reduce the synthesis of corresponding branched-chain esters (Frenkel and Patterson, 1973; Brackmann et al., 1993).

Processing, especially cutting of apples has a great impact on the alteration of VOC profiles. Some VOCs are only released due to cell disruption, resulting in an interaction previously separated enzymes and substrates (Buttery, 1993). Therefore, VOCs present in intact fruit tissue may be classified as primary compounds, whereas VOCs produced as a result of tissue disruption should be classified as secondary (Drawert et al., 1969). Tissue destructions lead to the release and mixing of vacuolar, cytoplasmic and nucleic enzymes and substrates normally sequestered in different cell compartments, enabling the activation of other biochemical pathways (Beaulieu, 2010). Additionally, cell membrane degradation following the processing led to the release of substrates available for VOC synthesis (see lipoxygenase pathway).

2.4.2 Biosynthetic pathways and related enzymes

Several pathways are involved in the synthesis of VOCs, which, themselves, comprise at least five chemical substance classes (Dixon and Hewett, 2000). The synthesis of VOCs by all these biosynthetic pathways requires specific primary precursors, i.e., usually fatty acids (membrane lipids), amino acids or carbohydrates. Therefore, the formation of VOCs is highly regulated during fruit development in terms of amount and composition of these precursors (Sanz et al., 1997; Song and Bangerth, 2003). According to Schaffer et al. (2007), four important pathways are mainly responsible for the synthesis of VOCs.

1. Straight-chain esters, alcohols and aldehydes from lipids, mainly linolenic and linoleic acids are synthesised via β-oxidation and lipoxygenase activity (Paillard, 1990; Rowan et al., 1999).

2. Branched-chain esters, alcohols and aldehydes are derived from isoleucine via the amino acid pathway (Tressl and Drawert, 1973; Hansen et al., 1993; Rowan et al., 1999; Matich and Rowan, 2007).

3. Terpenoids are produced via the mevalonate (Schwab et al., 2008; Sanz and Pérez, 2010) and the deoxyxylulose phosphate pathways (Eisenreich et al., 2004).

4. Phenylpropanoids are formed using the phenylpropanoid pathway (Gang et al., 2001).

In apples, however, only the fatty acid (β-oxidation and lipoxygenase) and the amino acid pathways are relevant and involved in the synthesis of esters and alcohols, which represent the main groups of aroma compounds (Dixon and Hewett, 2000). According to Dixon and Hewett (2000), the anaerobic fermentation pathway is additionally relevant for the formation of acetaldehyde, ethanol and resultant ethyl esters. The last step in the biosynthesis of all these esters is the esterification by alcohol acyltransferase (AAT), which depends on the availability of C2-C8 acids and alcohols (Paillard, 1979; De Pooter et al., 1981; Knee and Hatfield, 1981). According to Paillard (1979), the condensation of an alcohol and a carboxylic acid to volatile esters is depending on acyl-CoA (Fellman et al., 1991). Differences in ester profiles among cultivars may be attributed to the distinct substrate-specificity of AAT. This enzyme property may, in turn, be specific for fruit of a certain cultivar (Fellman et al., 2000). Esterases may hydrolyse acetate esters if these are available at high concentrations, thus balancing the concentrations of acetate esters and their corresponding alcohols (Goodenough, 1983; Fellman et al., 2000).

Anaerobic biosynthetic pathway

Very low O_2 concentrations prevent the normal oxidation of reduction equivalents, produced in glycolysis and tricarboxylic acid cycle (TCA), via the mitochondrial electron chain reactions and, thus, the TCA cycle activity inhibited. Under such conditions, generation of biochemical energy (i.e. ATP) to run the cellular metabolism is restricted to glycolysis, which, in turn, can only be active, when the re-oxidation of the thereby formed $NADH/H^+$ via the anaerobic pathways, especially that of ethanol formation, is activated.

In a first step, pyruvate decarboxylase (PDC) converts the pyruvate from the glycolysis to acetaldehyde and CO_2, leading to increase acetaldehyde concentrations (**Fig. 2.1**). In the final reductive step, alcohol dehydrogenase (ADH) reduces acetaldehyde to ethanol and, thus, oxidises the NADH/H^+ (Buchanan et al., 2015; Schopfer and Brennicke, 2016). Ethanol but also other alcohols are, however, direct precursors of esters (Dimick et al., 1983).

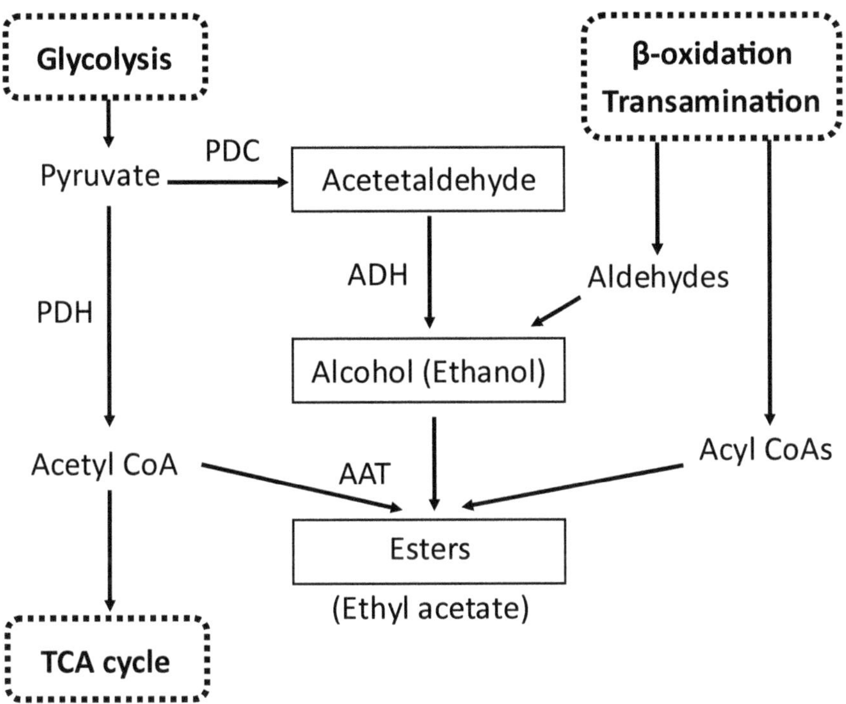

Figure 2.1. Anaerobic synthesis of acetaldehyde, ethanol, and esters (adapted from Mathews and van Hold, 1996). Highlighted text represents compounds that accumulate under hypoxic conditions. (PDH = pyruvate dehydrogenase, PDC = pyruvate decarboxylase, ADH = alcohol dehydrogenase, AAT = alcohol acyltransferase, TCA = tri-carboxylic acid.) (according to Dixon and Hewett, 2000)

Fatty Acids Pathway (β-Oxidation and Lipoxygenase)

In apples, straight-chain esters, alcohols and aldehydes are mainly synthesised via β-oxidation and lipoxygenase (LOX) enzymatic pathways (Sanz et al., 1997; Echeverría et al., 2004; Defilippi et al., 2005; Reineccius, 2006). While the β-oxidation pathway is prevalent in intact fruit, the lipoxygenase pathway (LOX) is activated after fruit tissue was disrupted (Schreier, 1984; Bartley, 1985; Gardner, 1995; Sanz et al., 1997; Rowan et al., 1999). The increase in lipid synthesis and membrane fluidity during fruit ripening also enhances the availability of fatty acids and increases the activity in LOX pathway (Wooltorton et al., 1965; Galliard, 1968; Guadagni et al., 1971; Bartley, 1985). In apples, the predominant precursors of the VOC synthesis via these pathways are fatty acids with 16 and 18 carbon atoms (C16:0, C18:0, C18:1, C18:2 and C18:3), mainly linoleic and linolenic acids (Galliard, 1968; Song and Bangerth, 2003; Reineccius, 2006).

β-Oxidation

The mechanisms of β-oxidation are well established (Goepfert and Poirier, 2007) and detailed described (e.g., Espino-Díaz et al., 2016; **Fig. 2.2**). In a first step, fatty acids are activated to their corresponding CoA-derivate by the acyl-CoA synthase. This synthase requires ATP, Mg^{2+} and coenzyme A with sulfhydryl functional group (CoASH). From this acyl-CoA/fatty acid, one C2 unit (acetyl-CoA) is removed in each cycle of β-oxidation and it is converted to shorter chain acyl-CoAs. This reduction requires again CoASH, flavin adenine dinucleotide (FAD), nicotinamide adenine dinucleotide (NAD) and H_2O. Acyl-CoA reductase reduced acyl-CoAs to aldehyde, which aldehyde can be further transformed to alcohol by the alcohol dehydrogenase (ADH) and afterwards to an ester by the alcohol acyl-CoA transferase (AAT). However, amino acids probably also provide substrates for acetate esters synthesis by β-oxidation (Rowan et al., 1997).

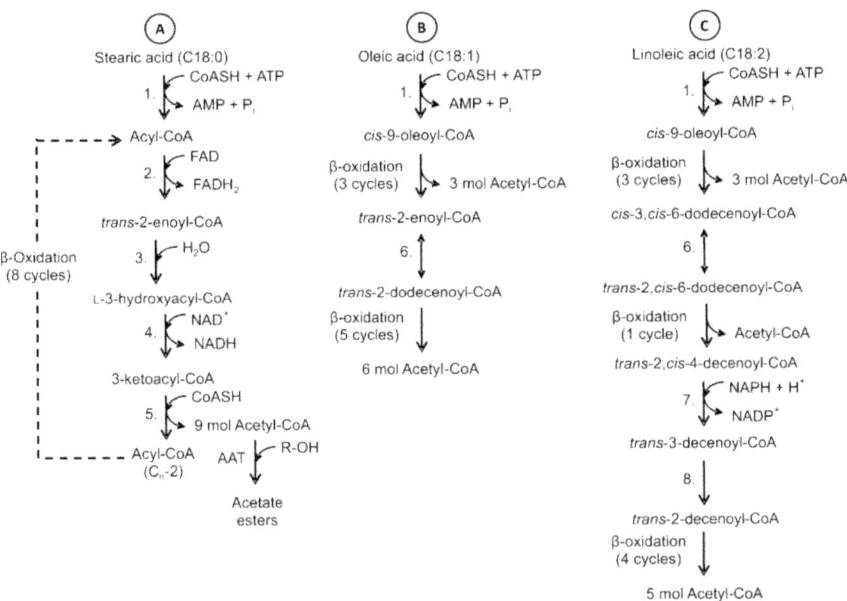

Figure 2.2. β-Oxidation pathway of C18 fatty acids. A) Saturated fatty acid (stearic), B) unsaturated fatty acid with one double bond in cis configuration (oleic), C) unsaturated fatty acid with two double bonds in cis configuration (linoleic). (adapted from Espino-Díaz et al., 2016)

Lipoxygenase (LOX) pathway

In the LOX pathway, polyunsaturated fatty acids are catalysed to saturated and unsaturated C6 and C9 aldehydes and alcohols (c.f. Espino-Díaz et al., 2016; **Fig. 2.3**). The required polyunsaturated fatty acids, primary linoleic or linolenic acid, are released from triacylglycerols, phospholipids or glycolipids by acyl hydrolases. They are further converted to 13-hydroperoxides by lipoxygenase and cleave to hexanal by hydroperoxide lyase (HPL). Analogous to β-oxidation, the thus generated aldehydes (e.g. hexanal) are transformed to alcohols (ADH) and further to esters (AAT).

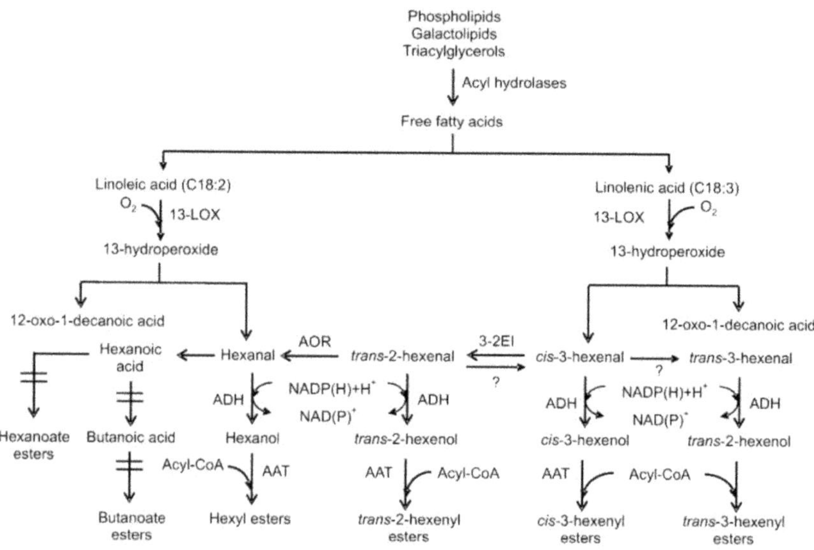

Figure 2.3. Lipoxygenase pathway in the production of acetate esters from the catabolism of linoleic and linolenic acids. LOX = lipoxygenase, HPL = hydroperoxide lyase, ADH = alcohol dehydrogenase, ALDH = aldehyde dehydrogenase, AOR = NADPH-dependent alkenal/one oxidoreductase, 3-2EI = cis-3: trans-2-enal isomerase, AAT = alcohol acyltransferase. (adapted from Espino-Díaz et al., 2016)

Amino Acid Pathway

In the amino acid pathway a broad number of compounds, including alcohols, carbonyls, acids and esters can metabolized. For the aroma biosynthesis in fruit, alanine, valine, leucine, isoleucine, phenylalanine and aspartic acid are direct precursors (Sanz et al., 1997; Baldwin et al., 2002). In apples, leucine, isoleucine and valine are the most important amino acids for the biosynthesis of branched-chain alcohols, carbonyls and esters (Sanz et al., 1997; Reineccius, 2005; Defilippi et al., 2009; Tzin and Galili, 2010). These amino acids are synthesized in the chloroplasts (Diebold et al., 2002; Schuster and Binder, 2005), where free amino acids originate from proteolysis (Smit et al., 2009). In this pathway, the activity and the selectivity of various enzymes are more relevant for the respective concentrations of branched-chain esters than is substrate availability (Rowan et al., 1997). In 'Braeburn', 'Granny Smith', 'Fuji', 'Red Delicious', and 'Royal Gala' apples, Rowan et al. (1997) found very distinct conversion ratios from amino acids to volatiles. However, different metabolic rates, in particular for leucine and isoleucine, were also observed (Rowan et al., 1997), which may explain the cultivar-specific characteristics in these VOCs. Espino-Díaz et al. (2016) evaluated the metabolisation of amino acids during the biosynthesis of VOCs. In **Fig. 2.4** this is in detail illustrated on the example of isoleucine. In the first step of the degradation of amino acids, the enzyme branched-chain amino transferase (BCAT) removes the amino group (Diebold et al., 2002; Schuster and Binder, 2005) thus forming the corresponding α-keto acid, in this case 2-oxo-3-methyl pentanoate (Dickinson et al., 2000; Graham, 2002; Liu et al., 2008; Smit et al., 2009). This degradation reaction, located in the mitochondria (Diebold et al., 2002; Schuster and Binder, 2005), may help to maintain the balance between NAD+ and NADH/H^+ (Espino-Díaz et al., 2016). The acceptor of the amino group, 2-oxo-ketoglutarate (Marilley and Casey, 2004; Ardö, 2006; Liu et al., 2008; Schwab et al., 2008), is then converted into glutamate by the glutamate dehydrogenase (GDH) (Schuster et al., 2006; Liu et al., 2008).

Via three different reactions (Marilley and Casey, 2004; Matich and Rowan, 2007; Liu et al., 2008), the α-keto acids can be further transformed into

(1) hydroxy acids (by a hydroxyacid dehydrogenase, HDH), which do not contribute to aroma;

(2) branched aldehydes (by an α-keto acid decarboxylase, e.g. PDC), which can be further converted into branched alcohol (ADH) or branched fatty acid (AldDH) and

(3) branched-chain acyl-CoA (by branched-chain α-keto acid dehydrogenase, BCKDH), which can be further converted into fatty acids (PAT) or branched-chain esters (AAT).

An additional pathway, in which 2-methyl-2-butenyl-CoA is degraded to acetyl-CoA and propyl-CoA through β-oxidation reactions (Graham, 2002), was also described (Espino-Díaz et al., 2016) but obviously not relevant for apples and therefore not shown here.

In the reactions, the corresponding α-keto acids where converted into corresponding branched aldehydes, branched alcohols or branched fatty acids (Ile = 2-methyl(-butanal/-butanol/-butanoic acid); Leu = 3-methyl(-butanal/-butanol/-butanoic acid); Val = 2-methyl(-propanal/-propanol/-propanoic acid)), as well as branched-chain acyl-CoA (Ile = 2-methyl butanoil-CoA; Leu = 4-methyl butyl-CoA; Val = 2-methyl propyl-CoA) and esters.

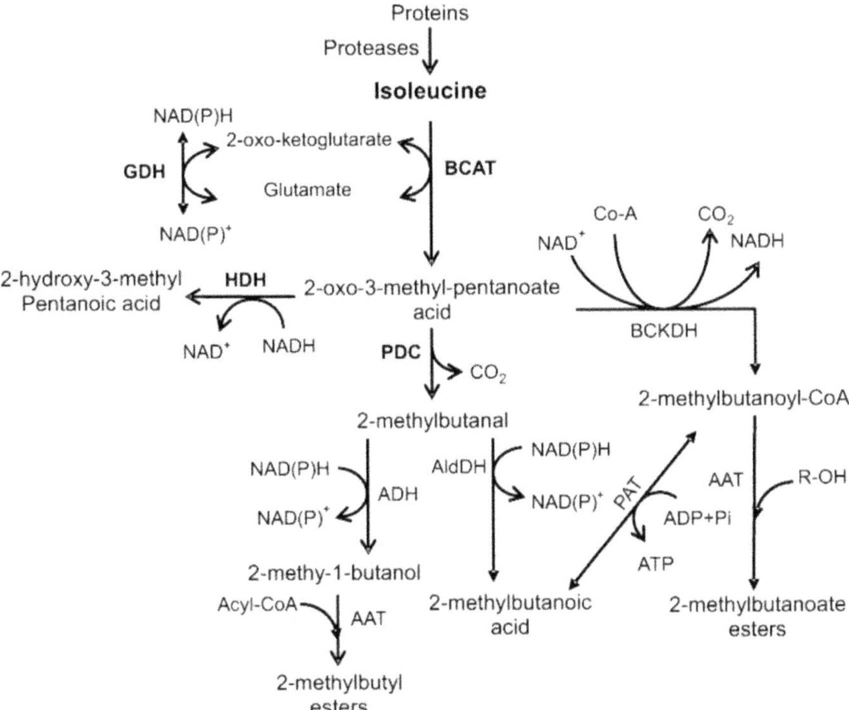

Figure 2.4. Biosynthesis of VOCs from isoleucine. BCAT = branched-chain amino acid transferase, GDH = glutamate dehydrogenase, PDC = pyruvate decarboxylase, BCKD = branched-chain α-ketoacid dehydrogenase, HDH = 2-hydroxy-3-methylpentanoic acid dehydrogenase, PAT = phosphate acetyltransferase, AldDH = aldehyde dehydrogenase, ADH = alcohol dehydrogenase, AAT = alcohol acyltransferase. (modified after Espino-Díaz et al., 2016)

2.5 Respiration and ethylene biosynthesis

The entire process of respiration involves a series of three interconnected metabolic pathways (glycolysis, tricarboxylic acid cycle and mitochondrial electron transport chain or end-oxidation) each of which is catalysed by a number of specific enzymes (Kadereit et al., 2014; Buchanan et al., 2015). Low respiration rates are assumed to indicate low metabolic activity, which is directly associated with longer shelf life (Saltveit et al., 2001). High respiration activity may often lead to increased secondary VOCs production (Beaulieu, 2006). Cutting and the concomitant metabolic wounding responses generally boost respiration (Buchanan et al., 2015). The intensification is only temporary and may decrease again within 24 h (Finnegan et al., 2013).

Decreased O_2 and/or increased CO_2 concentrations compared to that of the ambient air primarily slowed down the respiration processes of apple slices (Rocculi et al., 2006). The O_2 and the CO_2 concentrations inside the tissues of fresh-cut fruit slices are highly influenced by the headspace atmosphere and rapidly equilibrate to values similar to this due to the lack of a gas diffusion barrier (Soliva-Fortuny et al., 2001; Soliva-Fortuny et al., 2005). For fresh-cut fruit, the Low Oxygen Limit (LOL), below which anaerobic fermentation is induced, is largely reduced (Yearsley et al., 1996) and found to be about half that of intact unprocessed apples (Lakakul et al., 1999). This is attributed to both lower diffusion resistance and path length due to cutting and skin removal (Soliva-Fortuny et al., 2005). Soliva-Fortuny et al. (2005) reported that anaerobic respiratory processes were triggered only after three weeks of storage at 0 % O_2.

The metabolic wounding responses induced by cutting are also associated with the increase in ethylene emission (Soliva-Fortuny et al., 2002, 2005). Because the conversion of 1-amino-cyclopropane-1-carboxylic acid to ethylene inevitably requires O_2, very low O_2 concentrations may limit the ethylene synthesis (Yang, 1981). This has, e.g., been reported by Both et al. (2014) for intact 'Royal Gala' apples stored under ultra-low oxygen and by Soliva-Fortuny et al. (2005) for fresh-cut apple slices packed and stored at 0 % O_2 and other atmospheric compositions. At O_2 concentrations above these limits, the tissue ethylene content was independent of the O_2 availability. The storage of fresh-cut apple slices led to increased internal ethylene concentrations in all atmospheres. However, samples initially stored under anaerobic conditions of anoxia, nevertheless showed a slightly lower ethylene release after re-exposure to atmosphere (Soliva-Fortuny et al., 2005). Both synthesis of VOCs and ethylene are interconnected because ethylene is involved in some steps of several pathways of VOC syntheses, although ethylene does not regulate all aspects of flavour production (Schaffer et al., 2007).

2.6 Microbial spoilage and control strategies

The growth of microorganisms often represents the main limitation of fresh-cut fruits (Ragaert et al., 2011). It is directly related to the processing of ready-to-eat apples, which involves cutting and, if necessary, peeling. Thus, processing inevitably removes the essential physical and chemical barrier, i.e. the epidermis. This boundary tissue normally impedes microbes to establishing on the product surface (Martin-Belloso et al., 2006; Bhagwat, 2006). In contrast, if damaged, cell walls and epidermal wax layers facilitates microbes the attachment on the cut surfaces (Graça et al., 2015; Bhagwat, 2006). The cutting process also transfers microorganisms from the apple surface to the fruit flesh and largely increases the overall surface available to microbes for attaching and colonising (Bhagwat, 2006). In addition, microbial cross-contamination is highly possible at the various steps of the production chain of fresh-cut apples (Nguyen-the and Carlin, 1994). As a result, various microorganisms, including human pathogens (Bhagwat, 2006), which naturally attach to fruit surfaces during growth or during harvest, processing and handling (Beuchat, 1996; Heard, 1999) may be found on fresh-cut apples (Qadri et al., 2015). The unavoidably damage of tissue and cellular structures also leads to the leakage of nutrients and cellular fluids (Heard, 2002). Because these cell saps contain high concentrations of sugars and organic acids (Chen et al., 2016), they represent a fertile environment for microbial growth. Therefore, minimal processing strongly enhances microbial spoilage of fresh-cut produce if compared to intact apples (Busta et al., 2003).

In fresh-cut fruits, the inactivation or inhibition of a wide range of microorganisms, such as *Listeria monocytogenes*, *Salmonella* sp., sulfite-reducing *Clostridium*, *Escherichia coli*, *Enterobacteriaceae* sp., aerobic mesophilic bacteria, *Staphylococcus aureus*, yeasts and moulds are important to stabilize the produce to increase shelf life and to guarantee food safety (European Commission, 2005). On intact apples, fungi (*Penicillium, Rhizopus*, *Trichoderma, Cladosporium, Alternaria* and yeasts) are major microbes (Nguyen-the and Carlin 1994; Teixido et al., 1999). Yeasts in particular show a great diversity of genera (*Candida, Cryptococcus, Debaryomyces, Kloeckera, Kluyveromyces, Pichia, Rhodotorula, Saccharomyces* and *Zygosaccharomyces*; Tournas et al., 2006; Ruiz-Cruz et al., 2010), some of which are also associated with food spoilage (Graça et al., 2015). Various studies could not reveal any pathogenic microbes such as *Salmonella, E. coli, C. sakazakii* or *L. monocytogenes* on apples

(Abadias et al., 2006; Graça et al., 2015). Directly after cutting, yeasts and moulds are the dominant microbial groups (Soliva-Fortuny et al., 2004), but packaging and storage leads to a rapid transformation of this initial population by the replacement by gram-positive bacteria (Lanciotti et al., 1999; Lamikanra and Watson, 2000; Soliva-Fortuny et al., 2004). The German Association for Hygiene and Microbiology (DGHM 2011) defines and recommends maximal microbial loads, which are generally strictly adhered to by German producers. The total loads of anaerobic mesophilic bacteria (TAMB), yeasts and moulds should not exceed 7, 5 or 3 log CFU g^{-1}, respectively, at the end of shelf life. In addition, in the European Union, fresh-cut fruits must be completely free from *Salmonella* sp., *E. coli* and *L. monocytogenes*. Soliva-Fortuny et al. (2004) reported the initial counts of aerobic mesophilic microorganisms and yeast/moulds of apples after cutting with 2 log CFU g^{-1} and 1.7 - 2.2 log CFU g^{-1}, respectively.

The initial microbial population directly impacts the microbiological status of the finished product (Zhuang et al., 2003). Low initial loads, however, do not guarantee microbial stability during storage (Bhagwat, 2006). Because no current processing technique totally inactivates all adherent microorganisms without quality losses of fresh-cut fruit, best products require high-quality raw materials, stable and well-organized distribution and cooling conditions, and optimised logistics (Artes et al., 2007; Bhagwat, 2006). This underlined the necessity of Good Manufacturing Practice (GMP) and Hazard Analysis and Critical Control Points (HACCP) in fresh-cut fruit production (James et al., 2011). As fresh-cut fruit are consumed raw, heat treatments such as surface sanitation are possible. To extend the shelf life of fresh-cut apples by inhibition of microbial growth, refrigeration to temperatures below 4°C is essential (Smith et al., 1998).

Besides temperature, common approaches to control microorganisms include modified atmosphere packaging (MAP), washing with different sanitizing solutions, biocontrol (i.e. inhibition of bacterial growth by various antagonists), use of edible coatings or the application of non-thermal preservation methods (UV irradiation, cold atmospheric plasma). The use of preservatives is mostly abandoned due to the health-conscious of consumers. In the context of this study, modified atmosphere packaging and sanitiser solutions, which are common methods, will be described in more detail.

Packaging is a key instrument to maintain quality and extend shelf lives of fresh-cut products. Besides syrup immersion, MAP techniques typically combined with refrigerated storage are widely used to retard microbial spoilage (Heinrich et al., 2016; Li, et al., 2011; Montero-Calderon et al., 2008) and preserve food quality (Caleb et al., 2013). The composition of the modified atmosphere pronouncedly affects the development of microbial loads and quality. For the packaging design, the gas transmission rates of the packaging film and the respiration rate of the product(s) are of paramount importance for optimal packaging conditions (Conte et al., 2011). Soliva-Fortuny et al. (2004) pointed out that MA acts fungistatic but could not reduce initial loads. Low O_2 (1 - 5 %) concentrations can inhibit aerobic microorganisms growth (Farber, 1991, 2003; Al-Ati and Hotchkiss, 2003). Anaerobic conditions (<1 %), however, enhance the growth of anaerobic bacteria (Babic and Watada, 1996), including foodborne pathogens (e.g. *Clostridium botulinum*; Austin et al., 1998). High CO_2 concentrations act as an antimicrobial agent, with concentrations beyond 5 % being fungi- and bacteriostatic, inhibiting the growth of various spoilage microorganisms. According to Jay (1986), moulds and gram-negative bacteria are most sensitive to CO_2. This effect is mainly related to the reduction of pH due to the formation of carbonic acid from the CO_2 dissolved in cell sap, and its interference with the cellular metabolism (Brackett, 1997). In this context, the reduction of the maximum microbial growth rates and the population densities during the lag phase of microbial growth may result from the low pH (Devlieghere and Debevere, 2000). The ability of the increased CO_2 concentrations to inhibit microbial growth significantly depends on temperature, type and growth phase of microorganism, and on water-activity and chemical composition of the products (Devlieghere and Debevere, 2000).

Washing or sanitisation dippings are widely used to extend shelf life by reducing microbial loads and maintaining produce quality. Particularly acid dippings inhibit and control the microbial growth on fresh-cut fruit (Rahman et al., 2011). On citric acid-treated (0.5 %) 'Fuji' apple slices, bacterial loads were reduced by approx. 2.1 log CFU g^{-1} (Chen et al., 2016). The duration of the immersion in sanitisation/anti-browning dippings usually ranged from 1 – 5 min (Soliva-Fortuny and Martin-Belloso, 2003a). Washing or dipping also rinse the enzymes and substrates, released by cutting, from the product surface (Soliva-Fortuny and Martin-Belloso, 2003a). In addition, the treatments stabilise the fruit surface and delay physiological decay by preventing degradative processes (Soliva-Fortuny and Martin-Belloso, 2003a). Thus, washing or dipping represents important methods to remove pathogens during the processing of fresh-cut produce (Beuchat and Ryu; 1997; Burnett and Beuchat, 2001). However, the used water or solutions may also be a source of cross-contaminations (Bhagwat, 2006). To minimise this risk, water should not recirculate in processing lines and dippings should not reuse for multiple batches (Bhagwat, 2006). The contamination of sanitisation dippings with microorganisms, plant tissue and juices from previous batches also change their chemical compositions (decrease in conductivity; increases in soluble solid contents and osmolality; changes in pH) resulting in a decrease antibacterial effectiveness (Bhagwat, 2006).

Disinfectants available for washing and sanitisation have various efficacies and impacts on sensory attributes of the treated products (e.g. colour; Putnik et al., 2017). Various chemicals sanitisers including acids, chlorine, chlorine dioxide, hydrogen peroxide, ozone, bromine, iodine, trisodium phosphate and quaternary ammonium compounds were reviewed by the WHO (1998). Natural antimicrobials comprise citron essential oil, hexanal, 2-(E)-hexenal, citral and carvacrol, alone or in combination (Siroli et al., 2014), phenols and aldehydes (Soliva-Fortuny and Martin-Belloso, 2003a). Organic acids (lactic acid, citric acid, acetic acid, tartaric acid) and their salts represent commonly used natural organic compounds that are effective antimicrobial agents for fresh-cut fruit and highly accepted by consumers (Bari et al., 2005; Uyttendaele et al., 2004). Especially citric- and ascorbic acid can reduce microbial loads and prevent browning (Priepke et al., 1976; Soliva-Fortuny and Martin-Belloso, 2003a), based on pH reduction in the solution and within the cells, disruption of membrane transport and/or permeability, and anion accumulation (Beuchat, 2000). The acidification of the product surface was also recommended to inhibiting the growth of foodborne pathogens (Soliva-Fortuny and Martin-Belloso, 2003a), such as *L. innocua* (Karaibrahimoglu et al., 2004), *Salmonella* and *Shigella* (Escartin et al., 1989; Golden et al., 1993; Leverentz et al., 2001). Furthermore, a combination of MAP and the application of Ca-ascorbate or a mixture of ascorbic and citric acid reduced microbial spoilage of fresh-cut apples (Putnik et al., 2017).

3 Respiration and storage quality of fresh-cut apple slices immersed in sugar syrup and orange juice

Article I

Guido Rux [1], Oluwafemi J. Caleb [1,2], Antje Fröhling [1], Werner B. Herppich [1] and Pramod V. Mahajan [1,*]

[1] Department of Horticultural Engineering, Leibniz Institute for Agricultural Engineering and Bioeconomy (ATB), Max-Eyth-Allee 100, 14469, Potsdam, Germany

[2] Postharvest and Wine Technology Division, Agricultural Research Council, Stellenbosch, South Africa

* Correspondence: pmahajan@atb-potsdam.de

Received: 20 March 2017; Accepted: 3 August 2017; Published: 14 August 2017

Abstract: Storing fresh-cut apple slices in suitable fruit juice or sugar syrup is a general practice. However, application of this approach is mainly based on empirical knowledge, while systematic and comprehensive analyses of the relevant effects of this storage technique on keeping quality-related physiological properties of fresh-cut products is still missing. Hence, the aim of this study was to evaluate the impacts of complete immersion of fresh-cut apples in sugar syrup and fruit juice solution on respiratory behaviour and other relevant quality attributes (colour, tissue strength, and soluble solid and acidity). Sugar syrup and pure orange juice showed a high potential to store and protect fresh-cut apples. Results showed that only pure orange juice positively affected the produce quality by preventing browning effects. In addition, sugar syrup of 13.4–20 % most effectively prevented browning of apple slices and guaranteed high product quality retention during storage. The application of different liquid media provides a practical means to prevent browning and maintain product quality.

Keywords: fresh-cut fruit, respiration, minimal processing, quality

1. Introduction

The fresh-cut fruits sector has shown consistent growth in the last few years, due to changes in consumers' lifestyle and the growing demand for convenient, nutritional and safe "ready-to-eat" fruits (Van Rijswick, 2010; Baselice et al., 2015). This demand is further boosted by restaurants, fast-food outlets and institutional food-service operators seeking to reduce labour costs by buying minimally processed, ready-to-use or ready-to-eat fresh-cut fruits.

Fresh produce even after cutting remain physiologically active and respond to wounding. Hence, fresh-cut produce is different from intact fruit in terms of its physiology and postharvest handling requirements. Minimal processing of fresh-cut produce involves sorting, cleaning, washing, trimming, peeling, deseeding/coring, and cutting (such as chopping, slicing, shredding, chunking, and dicing). The first responses of fresh-cut produce to wounding include increase in both respiration and ethylene production (Mahajan et al, 2014). This has been largely studied for cut-vegetables showing that respiration rates of fresh-cut carrots can increase up to 3-fold directly after cutting (Iqbal et al., 2008). However, the increase is momentary, mainly due to initial stress response, as respiratory activity decreases to an equilibrium value within 24 h of processing (Finnegan et al., 2013). The loss of metabolic reserves in the commodity due to respiration processes means the hastening of senescence as the reserves that provide energy to maintain the commodity's living status get exhausted (Kader et al., 2003). This rapidly reduces the nutritional value (energy and nutritional values), and diminishes flavour quality, especially aroma compounds for the consumer and saleable mass for the producer. Therefore, higher respiration rates could indicate a more active metabolism and usually a faster deterioration. Thus, respiration is a key parameter for optimising postharvest treatments (Kader, 2010).

Mechanical injury of plant tissues caused by minimal processing intensifies water losses, thus, increasing the risk of wilting and dehydration (Toivonen and DeEll, 2002). This effect results from both artificially enhanced surface and from reduced diffusion resistance due to the removal of the natural protecting epidermal layer (Beaudry, 2000; Watkins, 2000). Fresh-cut processing also leads to enzymatic browning of cut surfaces as enzymes and substrates, normally sequestered in different cell compartments, become mixed with cytoplasmic and nucleic substrates and enzymes. Furthermore, minimal processing induces microbial growth due released exudates and eventually release of secondary fermentative metabolites (Artés et al., 2007).

Thus, various postharvest treatments have been investigated or applied over the years to minimize the adverse impact of minimal processing on fresh fruits and vegetables. These include the use of modified atmosphere and humidity packaging (Rux et al., 2015; Caleb et al., 2016), edible coatings (Rojas-Graü et al., 2007; Rojas-Graü et al., 2008; Salvia-Trujillo et al., 2015), thermal (e.g. Koukounaras et al., 2008) and plasma (e.g. Tappi et al., 2014) treatment, UV-C combined with MAP (López-Rubira et al., 2005) and sugar application (Verlinden and Garcia 2004; Nishikawa et al., 2005).

Postharvest sugar-syrup application or storing fresh-cuts in suitable fruit juice solutions as additive is a general practice. Sugars play an essential role in the complex metabolic processes of fresh produce as carbon and energy sources. Exogenous sucrose application has been reported to improve postharvest quality of broccoli by altering ethylene metabolic pathway (Nishikawa et al., 2005), and delay chlorophyll degradation (Coupe et al., 2003). Application of oxalic acid was shown to delay pericarp browning and maintain quality of 'Gola' litchi (Shafique et al., 2015). However, to the best of our knowledge, information on the physiological responses and changes in quality of fresh-cuts immersed in sugar solution or fruit juice is limited. The current industrial practices are based on "trial-and-error approach" without adequate measure of the dynamic response of fresh-cuts to sugar content or juice concentrations. This results in fresh-cut with shorter shelf life, poor quality, and consumers' rejection, which cumulates in enormous economic and postharvest losses.

This highlights the need for systematic and comprehensive scientific evaluation of these practices in order to provide relevant guiding tools. This will help in the optimization of the current industrial practices. Hence, the aim of the study was to investigate the impact of complete immersion of fresh-cut apples in sugar syrup and orange juice solution on respiratory behaviour and quality attributes. Special focus was laid on optimisation of the respective concentrations to provide best solutions for extending the shelf life of fresh-cut apples.

2. Materials and methods

2.1 Plant material

'Elstar' apples (*Malus domestica* Borkh.) were harvested at commercial maturity from an orchard located in Glindow, Germany (52°22'13.9"N 12°52'25.1"E) in 2015 and transported to the Department of Horticultural Engineering Laboratory, Leibniz Institute for Agricultural Engineering and Bioeconomy, Potsdam, Germany (ATB). On arrival, apples were stored at 1 °C for approx. 5 to 20 d until start of experiment. For thermal equilibrium, apples were transferred to the study temperature (13 °C), one day before the start of the experiments. This temperature was selected to simulate the average cool retail display condition obtainable. All experiments were performed in duplicate.

2.2 Fresh-cut preparation

Apples were processed under hygienic conditions at 13 °C. Cores were removed using a cork borer (24 mm) and each fruit was cut into 8 equally large slices. The slices were treated with NatureSeal FS (AgriCoat Natureseal Ltd., Great Shefford, UK) an antioxidant agent set to pH 2.45 for 5 min to prevent browning of cut surface. To ensure homogeneous distribution of samples, 1 slice from each of 8 slices per apple was placed in a tray. A total of 8 slices was placed in each polyethylene tray (190 mm x 115 mm x 75 mm) filled with different solutions (450 ml), and each treatment was replicated four times. Two replicates were stored together in a cuvette of gas exchange system. In this study, immersion solutions included normal air without any liquid (as positive control), deionized water (as negative control), glucose syrup and commercial orange juice. Freshly purchased syrup (invert sugar syrup = ⅓ glucose, ⅓ fructose, ⅓ sucrose) and orange juice (SSC = 9.8 % Brix; TA = 0.57 g 100 mL^{-1}) were mixed with deionized water (under clean conditions) to obtain solutions of different concentrations as common industrial practice (**Table 3.1**). The syrup solution containing 8 g L^{-1} OBST-SERVAL HC-2 browning inhibitor was obtained from a commercial supplier (Konserval Pharmacon Lebensmittelzusätze GmbH, Trittau, Germany). The initial mass of sliced apples and solution were recorded with an electronic balance (CPA1003S, Sartorius AG, Göttingen, Germany), and changes in mass were monitored throughout the storage duration.

Table 3.1. Different media used for dipping and storing of fresh-cut apple slices for 6 d.

Variant	Solute concentration
Control	Air
Water	Distilled water
Orange juice	25, 50, 100 %
Sugar syrup*	5, 10, 13.4, 20, 30 %

* Contained 8 g L^{-1} browning inhibitor

2.3 Respiration measurement

Due to the high temperature sensitivity of respiration, all experiments were conducted at a constant temperature of 13 °C for up to 6 d. For measuring respiration rates (RR) of apple samples immersed in different liquids and air during the storage, a custom-made closed gas exchange system was used (**Fig. 3.1**), consisting of 9 acrylic glass cuvettes (8.2 L volumetric capacity) fitted with GMP222 infrared CO_2 analysers (Vasalia, Helsinki, Finland). As replications, 2 trays containing apple slices were placed in each cuvette, the cuvettes hermetically closed with acrylic glass lid and the accumulation of CO_2 measured. These data were stored on a data logger (NetDAQ 2645A, Fluke Deutschland GmbH, Glottertal, Germany) and used to calculate RR. After regular CO_2 measurement intervals (330 min), cuvettes were automatically flushed with fresh air (0.25 L min^{-1}) via a pump (Laboport N811KN.18, KNF Neuberger GmbH, Freiburg, Germany) for 30 min to prevent excessive high CO_2 concentrations and, thus, allowing continuous measurements. After 6 d of measurements and storage, product quality was determined.

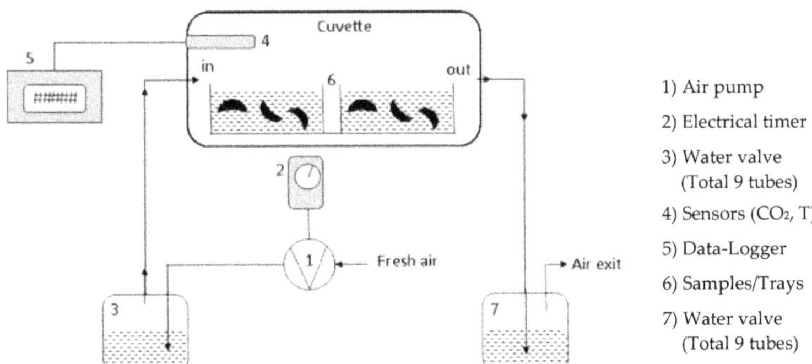

Figure. 3.1. Experimental setup for continuous measurement of respiration rate of fresh-cut apple slices.

In addition, the respiration rate of samples was measured after draining the liquid. In a closed glass container (4.2 L), rates of O_2 consumption (R_{O2}) and of CO_2 production (R_{CO2}) of the apple slices were monitored with O_2 (Lox-O_2 LuminOx, SST Sensing Ltd., Coatbridge, UK) and CO_2 sensor (Cozir, SST Sensing Ltd, Coatbridge, UK), respectively. R_{CO2} and R_{O2} were used to calculate the respiratory quotient RQ (RQ = R_{CO2}/R_{O2}).

2.4 Colour

For each treatment, surface colour of 8 apple slices was measured with a digital chroma-meter (CM-2600d, Konica Minolta Sensing Inc., Tokyo, Japan) and evaluated on the basis of Commission International del' Eclairage (CIE) colour system (L*, a*, and b*). Calibration of the CM-2600d was performed against a white and black tile, respectively. The colour parameter were analysed further as Chroma (C*; indicating the quantitative attribute of colour intensity), hue angle (h°; considered as the qualitative attribute of colour shades (0° = red-purple and 180° = bluish-green)), and browning Index (BI). The BI was calculated according to Maskan (2001):

$$BI = \frac{\frac{a^* + 1.75\,L^*}{5.645\,L^* + a^* - 3.012\,b^*} - 0.31}{0.17 \times 100}$$

2.5 Tissue strength

Tissue strength of apple slices was measured using a texture analyser (TA-XT Plus, Stable Micro Systems, Surrey, UK) fitted with a SMS-P/4 cylinder probe and set at 3 mm indentation to rupture tissues at a speed of 1 and 10 mm s^{-1} for test speed and post-test, respectively. Six samples per treatment were analysed and strength was expressed as maximum compression force (N).

2.6 Soluble solid content and total titratable acidity

From fresh tissue sap, obtained from the apple slices, soluble solid content (SSC) was measured using a hand refractometer (DR301-95, Krüss Optronic, Hamburg, Germany) and expressed as % Brix. The total titratable acidity (TA) of sap samples was measured potentiometrically by titration with 0.1 mol L^{-1} NaOH to end-point of pH 8.2 using an automated T50M Titrator with Rondo 20 sample changer (Mettler Toledo, Gießen, Germany). The TA concentration was expressed as g L^{-1} of malic acid fresh tissue sap. The ratio of "sugar to acid"-ratio of tissue sap was calculated and expressed as SSC/TA.

2.7 Microbial analysis

Microbial loads of drained apple slices were evaluated using total plate count method during the storage period. Total aerobic mesophilic bacterial count was determined using plate count agar (Carl Roth GmbH & Co. KG, Karlsruhe, Germany), while yeast and mould counts were determined using rose bengal chloramphenicol agar (Carl Roth GmbH & Co. KG). Four apple slices were taken from each treatment and transferred into sterile stomacher bags with 100 mL buffered peptone water. Samples were homogenized at speed 4 (10 strokes per sec) in a lab blender (BagMixer®400CC®, Interscience, Saint Nom, France) for 2 min and thereafter serial diluted by adding 30 µL of each diluent into 270 µL of peptone salt solution in Rotilabo®-microtest plates (96er U-profile, Carl Roth GmbH & Co KG). 100 µL from each dilution was pour-plated on respective growth media. Each treatment was analysed in triplicate (n = 3). Aerobic mesophilic bacterial count was enumerated after 72 h at 30 °C, and yeast and mould counts were enumerated after 5 d at 25 °C. The results were expressed as log colony forming unit per gram (log CFU g^{-1}).

2.8 Statistical analysis

Statistical analyses were carried out using Statistica Version 10.0 (StatSoft Inc., Tulsa, OK, USA). Data were subjected to analysis of variance (ANOVA) at 95 % confidence interval. Post-hoc test (Duncan Test) was used to test the statistical significant differences observed. All data were presented as mean ± standard deviation (SD).

3. Results and discussion

3.1 Respiration rates of apple slices immersed in a medium

Compared to that in air, respiration rates, i.e. the rates of CO_2 release, of apple slices were generally strongly reduced when samples were immersed in liquids, irrespective of the medium used (**Fig. 3.2**). This was due to the additional gas diffusion barrier and, hence, increased resistance to gas diffusion, generally reducing gas exchange from tissues to ambient (von Willert et al., 1995). This also diminished the O_2 availability at the apple slices and, therefore, slowed down respiration activity. In addition, the physical solubility of CO_2 in liquid and its possible chemical conversion into HCO_3^- initially impair the CO_2 release into cuvette headspace. For instance, the CO_2 release from the media into the headspace decreased by 53.0 to 65.4 % compared to that of controls.

All treatments significantly influenced ($p \leq 0.05$) CO_2 release into the headspace although following varying dynamics (**Fig. 3.2**). CO_2 release rates ranged from 19.7 to 69.0 mg kg^{-1} h^{-1} after 6 d of storage for all combinations tested. In controls, CO_2 release rates increased from initially 20.9 mg kg^{-1} h^{-1} for fresh cut apple slices to 48.1 mg kg^{-1} h^{-1} after 6 d of storage. For samples stored in water, the CO_2 release rates on day 6 increased by 31.3 % to a final rate of 63.2 mg kg^{-1} h^{-1}. These changes were comparable to those of the controls.

Varying the sugar concentration of syrups had significant impacts on the CO_2 release into headspace ($p < 0.05$). In 10 % sugar solutions, respiration of samples declined after 6 d of storage compared to controls. Slices stored in 5 % sugar syrup showed the highest CO_2 release rates (45.3 mg kg^{-1} h^{-1}) of all sugar syrup-stored samples and was not significantly different compared to control. Storage in sugar concentrations between 5 and 20 % differentially affected the respiration of samples. Increasing sugar concentrations of syrups to 20 or 30 % largely reduced CO_2 release rates of apple slices to 21.0 or 19.7 mg kg^{-1} h^{-1}, respectively, after 6 d.

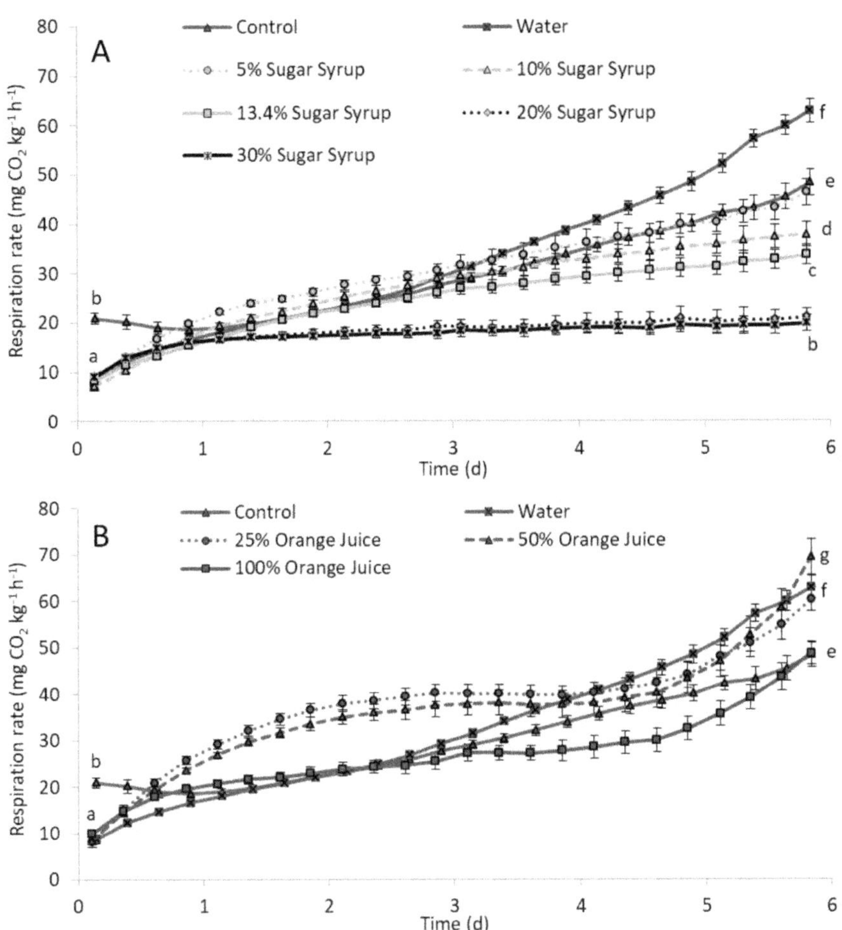

Figure. 3.2. Change in respiration rates of fresh-cut apple slices immersed in **sugar syrup (A)** and **orange juice (B)** with different concentrations

Compared to controls, storage in pure orange juice reduced respiration activity of samples from days 3 to 6, but significantly intensified it again to the same rate as found for controls on the last day. In contrast, samples immersed in 25 and 50 % orange juice showed higher rates of CO_2 release than controls during the entire test period and a similar trend after day 5, when CO_2 release rapidly increased. This could be due to the high microbial load that actively contributed in CO_2 production.

In diluted orange juice and in water, CO_2 release rates of apple slices increased on day 6 to 59.4 and 69.03 mg kg^{-1} h^{-1}, respectively. After 6 d of storage, samples immersed in 50 % orange juice showed the highest CO_2 release of all tested treatments. The rapid increase of CO_2 release at the end of storage may be explained by an accelerated growth of microorganism on the surface of the orange juice, which was visually observable on the last day of storage. Consequently, increases in CO_2 release resulted from both the enhanced respiration activity of apple slices and from that of the growing microorganism (see section 3.5). To evaluate the effect of orange juice on RR, it is necessary to analyse and compare the CO_2 release of apple slices after draining from liquid.

The effects of immerging apple slices in syrups on respiration activity of samples may closely reflect that of the well-studied MA-storage conditions. Gil et al. (1998), for example, showed that respiration of 'Fuji' apple slices was very low when stored under MA-conditions at 10 °C for up to 16 d but rapidly increased when slices were exposed to normal air.

The average RQ for aerobic respiration of fruits and vegetables under atmospheric conditions ranges between 0.7 and 1.3 (Kader et al., 2003). It is known since long that the RQ for fresh produce equals 1 when carbohydrates are used as respiratory substrates, while a RQ > 1 indicates the consumption of organic acidic (Fonseca et al., 2002). On the other hand, a rapid increase of RQ beyond 1 in plants also denotes the switch of aerobic respiration to anaerobic fermentation processes i.e. ethanol metabolism (Kader 1987). The extent of anaerobic metabolism is determined by the O_2 partial pressure available for fruits and vegetables (Beaudry 2000).

All treatments significantly ($p \leq 0.05$) influenced the RQ of apple slices. RQ's measured after 6 days of storage, ranged from 1.32 to 3.61 across samples of all combinations tested (**Fig. 3.3**). CO_2 release generally increased after taking samples out of the liquids, which, in turn, led the RQ increase from 0.89 to 1.32. This mainly results from the pronounced decreased in diffusion resistance in air compared to that in the liquids. As expected, rates of CO_2 release and O_2 uptake and, thus, the low RQ, remained unchanged in air-stored controls. This corresponds with findings of Conesa et al. (2007), showing that a higher RQ is associated with the increase in CO_2 production after cutting of fresh bell peppers.

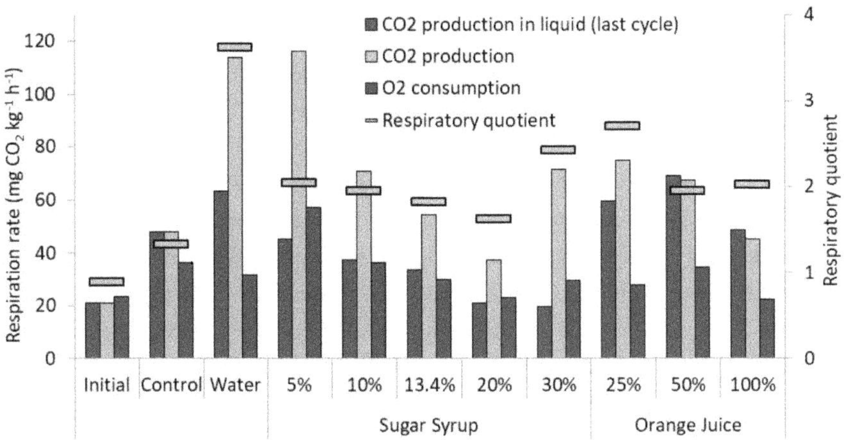

Figure. 3.3. CO_2 production, O_2 consumption and respiratory quotient of fresh-cut apple slices after draining liquid and 6 d of storage.

Draining of samples beforehand stored in water, showed the highest RQ (3.61) of all others tested. This was, again, due to the highest increase of CO_2 release in these samples. After storage in syrup of 5 % sugar content samples showed higher CO_2 release. Here, however, rates of O_2 consumption concomitantly increased, which, in turn, resulted in a pronouncedly lower RQ (2.04) than that found for water-stored apple slices. Increasing the sugar concentrations of syrups from 5 to 20 %, led the RQ decrease from 2.04 to 1.62. In contrast, if stored in syrup of 30 % sugar content, RQ of samples again, increased to 2.42, due to strong increase of CO_2 production. Then, the calculated RQ was the highest of all samples immersed in syrups of lower sugar concentrations (i.e. < 30 %). The strong increase of CO_2 production could be attributed due to two reasons. Firstly, the apple slices lost an average 10 % of mass due to osmotic dehydration. In turn, this reduced the respiration rate by 10 % to 64.2 mg kg^{-1} h^{-1} and RQ reduced to 2.2. Secondly, the osmotic dehydration might have increased the respiration rate of apple slices. This is in agreement with Castelló et al. (2009) who reported reduction in respiration rate of osmo-dehydrated and vacuum pulse treated apple slices at 10 °C.

For the orange juice treatment, CO_2 release measured after draining of samples decreased with increasing the orange juice concentration. However, in contrast to all other treatments, rates of CO_2 release observed after draining were smaller for samples of 50 and 100 % orange juice treatments than those measured for immersed slices (**Fig. 3.3**). CO_2 release rates of the latter were, in turn, higher than those found for samples immersed in sugar syrups. This may be due to the respiration of microorganism visible on surfaces of all orange juices before draining. The small (25 % orange juice) or very little (50 and 100 % orange juice) change in CO_2 production upon exposure to air, may point to a major effect of the high microbial load of these media, actively contributing to the CO_2 release. This means that the real CO_2 release of the slices are masked by the additional CO_2 release of the microorganisms. This real CO_2 release would be less without the microbial load and would also rise significantly after the removal (as with the other samples). The RQ obtained for apple slices stored in 50 and 100 % orange juice (1.94 and 2.03, respectively) was in the same range as that found for samples stored in 10 and 5 % sugar syrup solution. The RQ of 2.71 calculated for apple slices immersed in highly diluted orange juice (25 %) was highest next to that of water-stored samples, which was the highest of all tested liquids. The results of the RQ presented here highlighted that samples stored under all tested conditions more or less suffer from anaerobiosis at end of storage.

3.2. Colour

Surface colour of fresh-cuts and its changes during storage is one of the quality parameters, which have greatest impact on the purchase decision of consumers (Toivonen and Brummel 2008). In control apple slices, stored in air, the browning index (BI) significantly increased from the initial value of 31.3 to 57.3 after 6 d (**Table 3.2**). Immersion of sliced apples in syrup sugar had a significant impact ($p < 0.05$) on BI. Storing of samples in syrups of all sugar concentrations completely prevented browning of tissue surface. On the other hand, fresh-cut apples showed clear browning effects when stored in orange juice solutions of 50 and 25 % juice concentration but not in pure orange juice (100 %). In samples stored in water, tissue browning was very severe with the BI increasing up to 441.6.

Table 3.2. Colour measurement parameters and browning index (BI) of fresh-cut apple slices stored in different liquid media

Variant	**Colour**					
	*L**	*a**	*b**	*C**	*h°*	BI
Initial	85.0 ± 2.3 (a)	-1.0 ± 0.6 (a)	23.9 ± 2.1 (abc)	23.9 ± 2.1 (ab)	92.5 ± 1.4 (a)	31.3
Control	74.4 ± 3.1 (d)	4.6 ± 1.7 (c)	31.1 ± 3.4 (e)	31.5 ± 3.5 (e)	81.6 ± 2.7 (c)	57.3
Water	44.2 ± 8.3 (f)	19.5 ± 5.9 (e)	59.0 ± 4.7 (f)	62.4 ± 5.2 (f)	71.9 ± 5.3 (d)	441.6
Sugar syrup*						
5 %	81.0 ± 0.4 (bc)	-1.3 ± 0.0 (a)	21.3 ± 0.2 (a)	21.4 ± 0.2 (a)	93.5 ± 0.0 (a)	28.6
10 %	80.8 ± 1.3 (bc)	-1.2 ± 0.1 (a)	23.3 ± 0.7 (abc)	23.3 ± 0.7 (ab)	93.0 ± 0.1 (a)	32.0
13.4 %	82.6 ± 1.3 (bc)	-0.9 ± 0.3 (a)	21.7 ± 1.4 (ab)	21.8 ± 1.4 (ab)	92.2 ± 0.9 (a)	29.0
20 %	83.3 ± 2.3 (b)	-1.3 ± 0.3 (a)	22.8 ± 1.3 (ab)	22.9 ± 1.3 (ab)	93.5 ± 0.9 (a)	30.0
30 %	80.4 ± 0.8 (c)	-1.0 ± 0.1 (a)	25.2 ± 0.6 (bcd)	25.2 ± 0.6 (bcd)	92.3 ± 0.2 (a)	35.7
Orange juice						
25 %	55.8 ± 2.9 (e)	8.6 ± 1.4 (d)	26.3 ± 2.2 (cd)	27.8 ± 2.4 (d)	72.7 ± 2.2 (a)	73.5
50 %	73.0 ± 1.0 (d)	2.1 ± 0.3 (b)	26.9 ± 0.8 (d)	27.1 ± 0.9 (cd)	86.0 ± 0.3 (b)	46.8
100 %	82.2 ± 3.3 (bc)	-1.7 ± 0.6 (a)	24.3 ± 1.8 (bcd)	24.4 ± 1.8 (bc)	93.9 ± 1.4 (d)	32.6

Values are means ± standard deviation (n = 10). Different lower case superscript letters are significantly different ($p < 0.05$).

* Contained 8 g L^{-1} browning inhibitor

Browning of fruit tissues is strongly associated with the activity of polyphenol oxidase and by the concentrations phenolic compounds, but also by cell sap pH, tissue temperature and oxygen content of the tissues (Martinez and Whitaker 1995). Immerging fresh-cut products in adequate liquids largely reduces the oxygen concentration in tissues and, thus, helps preventing browning. This has been convincingly proven for apple slices stored in sugar syrup in this study.

Similarly, storing 'Golden Delicious' apple slices in 100 % N_2 prevented a decline of L^* values in contrast to exposure of samples to air (Soliva-Fortuny et al., 2002). In other studies on apple slices kept at 0-1 kPa O_2 after 7 d of storage (Gunes et al., 2001) or 5 kPa O_2 + 5 kPa CO_2 + 90 kPa N_2 (Rocculi et al., 2004), hue angle decreased and lightness values L^* increased. Furthermore, the use of anti-browning agent such as orange juice with additional citric acid added or commercial sugar solutions effectively prevented browning of apple slices in the present study. Consistently, Cocci et al. (2006) showed that combinations of anti-browning dipping (ascorbic and citric acids) and MAP successfully maintained colour attributes of apple slices stored at 4 °C for 8 d. Similarly, Cortellino et al. (2013) analysed the combination of dipping in citric and ascorbic acid with conventional and modified atmosphere on 'Golden Delicious' apple slices at 4 °C for 11 d, indicating that apple slices without dipping developed browning within the first 24 h irrespective of other storage conditions. The vacuum impregnation of 'Jonagold' apple slices with 10-40 % sugar solutions resulted in a decrease in lightness L^* at higher sucrose concentrations, even in the presence of acids (Biegańska-Marecik and Czapski 2007).

The above observation contrasts the results of the presented study. This difference in lightness L^* could be due to different hydration levels of impregnated and of liquid stored samples. Also, the accelerated browning effects observed on slices stored in 25 % orange juice solution and pure water could be due to the low antioxidant concentration of e.g. ascorbic acid in these liquids. It may also be attributed to their low concentration of organic acids, which resulted in higher pH around the apple slices and, thus, higher PPO activity (Weemaes et al., 1998). It has been shown that solutions of acidic pH or the addition of acidulants may effectively inhibit or control enzymatic browning (Nicolas et al., 1994; Whitaker 1994; Chiabrando and Giacalone 2012).

3.3 Tissue strength

Another important quality parameter evaluated in this study was tissue strength. Softening of fruit can be attributed to maturation, ripening and/or early degradation of tissues. Tissue strength is determined by cell size, biochemical and biophysical cell wall properties, cell-to-cell adhesion and tissue turgor (Toivonen and Brummel 2008). Tissue strength of freshly cut apple slices was 2.04 N, and it declined to 1.82 N in controls after storage (**Fig. 3.4**). In general, immersion of apple slices in liquids significantly affected ($p < 0.05$) tissue strength (**Fig. 3.4**). The effect, however, was largest in samples stored in pure water. Here, tissue strength declined by 63.9 % to 0.74 N at the end of storage. The pronounced impact of pure water might be due its low osmolarity, which probably facilitates water uptake by samples, resulting in hyper-turgidity and, eventually, cell lysis in mature apple tissues. This may have led to ruptured cell wall and weakened cell-to-cell adhesion, and, in turn, reduced tissue strength. This was also indicated by clear symptoms of cell disintegration after the end of the experiments.

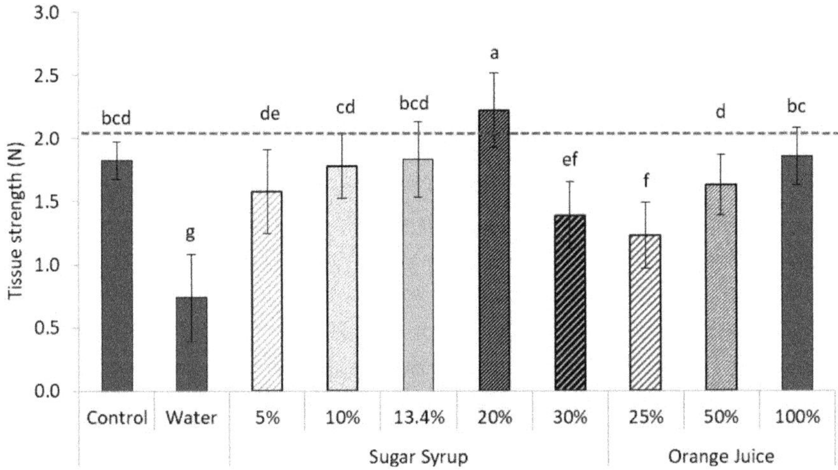

Figure. 3.4. Tissue strength of fresh-cut apple slices after 6 d storage in different liquids compare to initial quality (red dotted line). Error bar represents standard deviation of mean values (n = 8) at 95 % confident interval.

On the other hand, immersion of sliced apples in sugar syrup of various concentrations minimized or prevented softening (**Fig. 3.4**). Immersion of samples in syrup of 20 % sugar content even increased tissue strength by 8.9 %, compare to the strength of freshly cut slices. However, further increase of sugar content to 30 % led the sliced apple tissue strength decline to 1.39 N due to osmotically reduced tissue turgor, however, without any symptoms of cell disintegration as observed in water-treated samples. A similar response was observed for samples stored in orange juice solutions (25 – 100 %). Immersion of samples in pure orange juice and in 13.4 % sugar syrup solution reduced tissue strength to approx. the same degree. Previous studies showed that physical and chemical changes may affect textural integrity and enzymatic hydrolysis of cell wall pectic substances (Van Buggenhout et al., 2009). Hydrolysis of protopectins to water soluble pectins, diffusion of sugar into intercellular spaces, the decrease in cellulose crystallinity, ion movement from the cell wall and thinning of cell walls are assumed to intensify softening (Toivonen and Brummel 2008). Ponting et al. (1972) and Rojas-Graü et al., 2007 showed a significant reduction of firmness for ascorbic acid treated apple slices. In general, the current study showed that the osmolarity of the immersion media had the most pronounced influence on softening of apple slices.

3.4 Soluble solid content and total titratable acidity

Storing controls in air for 6 d did not significantly affect their soluble solid content (**Fig. 3.5**). In contrast, storage in pure water for the same time pronouncedly reduced the SSC of slices by about 58.5 %. In samples immersed in sugar syrups or orange juice solutions, SSC gradually changed according to the osmotic content of the respective solutions. If sample were stored in hypo-osmotic solutions their SSC declined. Storage in hyper-osmotic solutions consequently resulted in an increased SSC of apple slices (**Fig. 3.5**). To the best of our knowledge no other studies comprehensively evaluated the effects of storing apple slices in liquid medium on tissue sugar content. Similarly to the results obtained for controls in the presented study, in 'Jonagold' apple slices pre-treated with sucrose prior to vacuum packaging, SSC increased more pronouncedly during storage the higher the sucrose concentrations during pre-treatment were (Biegańska-Marecik and Czapski 2007). In 'Golden Delicious' apple slices stored in low oxygen at 4 °C for 12 d, the SSC of samples did not substantially change during the study period (Rocculi et al., 2004).

In contrast to SSC, the initial total titratable acidity (TA) of fresh-cut apple slices slightly declined in air-stored controls (**Fig. 3.5**). Storage in pure water, again, had the largest effect, reducing TA of respective samples to only 0.09 g 100mL^{-1}. Immersion in sugar syrup solutions reduced TA of apples slices by 19.6 to 24.5 %, irrespective of the syrup sugar content, except for 30 % sugar concentration, where a higher reduction (38.2 %) was measured. Although the acid concentration in the sugar solution was kept constant, the pronounced dehydration of apple slices and thus, dilution of the medium found at 30 % sugar might have induced diffusion of organic acids from the sliced apple tissues into the surrounding medium. In minimally processed apples stored under MA at 4 °C for 8 d Cocci et al. (2006) found no major change in TA content. However, TA of apple slices dipped (3 min) in antioxidant solution (1 % ascorbic acid plus 1 % citric acid) increased 20-fold prior of storage, most probably due to the uptake of the anti-browning solution. TA of samples treated with pure orange juice remained constant due to the high acidity of the juice; whereas immersion in diluted orange juices let the TA of the apple slices gradually decline by 43.0 %.

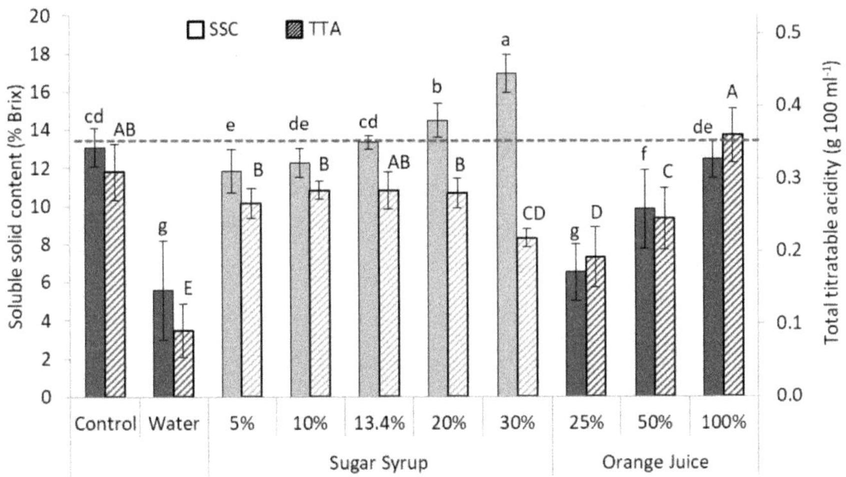

Figure. 3.5. Soluble solid content (SSC) and total titratable acidity (TA) of fresh cut apple slices after 6 d storage in different liquids compare to initial quality (red dotted line). Error bar represents standard deviation of mean values (n = 3) at 95 % confident interval.

3.5 Microbial analysis

Total aerobic mesophilic bacteria counts of freshly cut apple slices before immersion in the different liquids were 2.04 log CFU g^{-1} that of yeasts and moulds were 2.09 log CFU g^{-1} and 1.84 log CFU g^{-1}, respectively (**Fig. 3.6**). Counts of total aerobic mesophiles, yeasts and moulds increased in controls during air-storage. In sugar syrup stored apple slices, microbial growth remained unchanged from day 0 to day 3. Generally, microbial loads showed an increasing trend from day 3 to day 6, however further evidence is required with more number of samples to confirm these observations. Treatments with sugar syrups resulted in similar microbial counts for yeasts and moulds during storage. Increasing the sugar content to 20 % slightly reduced moulds compared to controls. Samples treated with 5 and 10 % sugar content had the highest total aerobic mesophilic bacteria count. Varying the orange juice concentration did not influence microbial counts. Nevertheless, microbial loads of samples, immersed in 25 to 50 % orange juice, were higher than on those stored in pure juice.

Generally, no treatment tested could fully prevent or even pronouncedly minimize microbiological growth. Yeasts, moulds and lactic acid bacteria are microorganisms associated with the spoilage of orange juice (Lawlor et al., 2009). These microorganisms are capable of growing at the low pH (pH 3.5–3.9) of orange juice (Lawlor et al., 2009). This may be one reason for the slightly higher microbial load in samples stored in orange juice. The German Association for Hygiene and Microbiology (DGHM 2011) defined standard value for mesophilic bacteria, yeasts and moulds (7, 5 and 3 log CFU g^{-1}, respectively) on fresh-cut and packaged fruits. Compared to these standard values, only fresh-cut apples stored in 5 % sugar solution exceeded the standard value for mesophilic bacteria counts.

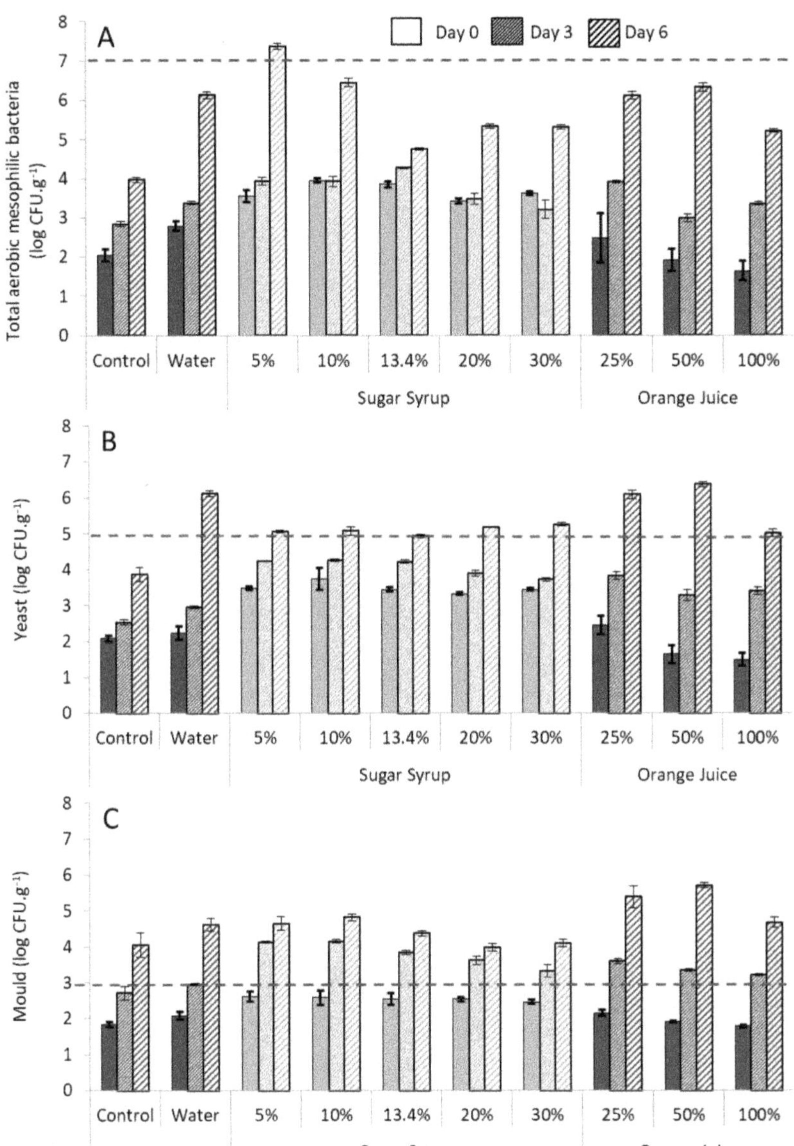

Figure. 3.6. Microbial analysis of fresh-cut apple slices stored in different liquids: Total aerobic mesophilic bacteria **(A)**, yeast **(B)**, and mould **(C).** The red dotted lines represent the standard values defined by the German Association for Hygiene and Microbiology (DGHM, 2011). Error bars represent standard deviation of mean values (n = 3) at 95 % confidence interval.

4. Conclusions

Sugar syrup and pure orange juice showed a high potential to store and protect fresh-cut apples. Results showed that only pure orange juice positively affected the produce quality by preventing browning effects. Storage of fresh-cut apples in sugar syrup and pure orange juice created modified conditions (absence of air), which slowed down physiological processes. Generally, no treatment tested could fully prevent or even pronouncedly minimize microbiological growth. Results indicated that CO_2 release from apple slices significantly declined depending on the sugar concentrations. Also, browning of samples was considerably reduced due to the absence of O_2 under these conditions, which, on the other hand, increased the risk of anaerobic respiration. Sugar content of 13.4 - 20 % showed the best results with smallest browning index and minimal rates of CO_2 release. In this range of sugar concentrations, respiration activity was considerably reduced, and tissue strength and soluble solid content are retained almost at the initial quality. The decrease in acidity may be balanced by adding ascorbic acid to the sugar solution. Increasing the sugar content to 20 % partially reduces the growth of microorganisms on the product surface.

In contrast, sugar solutions at concentrations below 13.4 % and diluted orange juice resulted in progressive reduction of produce quality. These concentrations were not able to reduce the microbial counts for yeasts and moulds during storage and showed the highest total aerobic mesophilic bacteria count. The standard value for mesophilic bacteria counts are exceeded for 5 % sugar solution. The use of diluted orange juice resulted in lower quality of samples compare to those stored in air.

4 Effects of modified atmosphere and sugar immersion on physiology and quality of fresh-cut 'Braeburn' apples

Article II

Guido Rux [1,2,*], Kimberly Bohne [1, 2], Christian Ulrichs [2], Susanne Huyskens-Keil [2], Karin Hassenberg [1] and Werner B. Herppich [1]

[1] Department of Horticultural Engineering, Leibniz Institute for Agricultural Engineering and Bioeconomy (ATB), Max-Eyth-Allee 100, 14469, Potsdam, Germany

[2] Division Urban Plant Ecophysiology, Section Quality Dynamics/Postharvest Physiology, Humboldt-Universität zu Berlin, Lentzeallee 55/57, 14195 Berlin, Germany

* Correspondence: grux@atb-potsdam.de; Tel.: +49-331-5599-920

Submitted: 30 Januar 2020

Abstract: Fresh-cut fruits are highly perishable and show a short shelf life. Therefore, storage in sugar syrup is commercially used to delay undesirable metabolic changes and to prevent browning. Investigations on the effects of syrup application on volatile organic compounds (VOCs), important for the consumers' sensorial acceptance, are rare. Low O_2 availability, as found in fruit slices after immersion, is probably one of the key factors. In the present study, fresh-cut 'Braeburn' apples were comparatively stored in air, in three modified atmospheres (O_2: 10, 5 and approx. 0 %; CO_2: 10, 15 and 20 %) or in sugar syrup (20 %). Relevant quality, physiological and microbiological parameters were evaluated on days 3, 6 and 10 of storage. Particular focus was laid on the evaluation of VOCs emission to indicate biosynthetic responses that affect aroma-relevant volatiles. Atmospheric storage of apple slices, among others, led to increased emissions of ethyl acetate, which may create an off-odour and negatively affect the customers' acceptance. In contrast, syrup-immersion of apple slices resulted in a pronounced loss of aroma but prevented the occurrence of off-odour. The results indicated metabolic changes, which were independent of O_2 availability. High CO_2 or VOCs concentrations in the fruit tissue may inhibit esterification or may induce a feedback inhibition of VOCs synthesis.

Keywords: minimal processing; sugar syrup storage; microbial analyses; volatile organic compounds; chemical prevention; ready-to-eat fruit salads

1. Introduction

Due to the inevitable wounding and tissue destructions by minimal processing, fresh-cut fruits become highly perishable and show a short shelf life. In addition, peeling or slicing enlarge the surface and lead to the release and the mixing of vacuolar, cytoplasmic and nucleic enzymes and substrates (Beaulieu, 2010). This facilitates the spread and growth of naturally occurring microorganisms, normally only located on the fruit skin (Nicoli et al., 1994). Processing may increase respiration, and the biosynthesis of ethylene and secondary metabolites; it may also activate enzymatic systems accelerating cell membrane degradation (Nicoli et al., 1994; Toivonen and DeEll, 2002). Besides microbial loads, rapid oxidative tissue browning and aroma loss are important factors limiting the customers' acceptance (Dixon and Hewett, 2000; Toivonen and DeEll, 2002).

In general, chemicals (Buta et al., 1999) and/or modified atmospheres (MA) during storage (Gorny et al., 1998) are used to control oxidative browning and to delay undesirable biological and biochemical changes (Beaulieu, 2010). Another approach, frequently applied in practise for bulk purchaser, is the storage of fresh-cut fruits in sugar syrup (Alzamora et al., 1993). For the first time, Rux et al. (2017) indicated that optimised storage of apple slices in 20 % sugar syrup successfully prevented enzymatic and oxidative browning and strongly reduced respiration, thus considerably extending shelf life. Rux et al. (2017) pointed out that especially the reduced gas diffusion lowered respiration and also rapidly diminished the O_2 availability resulting in conditions that closely reflect MA storage. However, further information on the effects of syrup application on physiology and quality of fresh-cut fruit is rare (Rux et al., 2019b).

This also relates to their potential synergistic interactive effects on the emission of volatile organic compounds (VOCs) of the immersed apple slices. In apples, a complex mixture of more than 300 VOCs (Dixon and Hewett, 2000), including esters, aldehydes, ketones, alcohols and hydrocarbons (Flath et al., 1967; Espino-Díaz et al., 2016) is responsible for the aroma attributes (Beaulieu, 2010) and is, therefore, important for the consumers' sensorial acceptance (Beaulieu, 2010). VOCs may occur as permanent flavour components (Espino-Díaz et al., 2016) or are elicited as secondary components by enzymatic actions closely attributed to processing.

Biosynthesis of VOCs utilises numerous pathways, which involve different enzymes and substrates. These pathways are all vastly affected by various interactions such as respiration and ethylene synthesis (Yabumoto et al., 1977; Sanz et al., 1997; Fellman et al., 2000; Espino-Díaz et al., 2016). For instance, low O_2 concentrations, as found after immersion, may negatively affect flavour and aroma (El Hadi et al., 2013). Multitude of biochemical reactions may develop VOCs-based off-flavours during storage, which affect quality and shelf life and must be minimised. Currently, knowledge of the impact of sugar solution on the aroma development of immersed fresh-cuts is very limited.

The present study evaluates quality-relevant physiological and microbiological parameters to elucidate the metabolic behaviour of fresh-cut 'Braeburn' apples immersed in sugar syrup. Of particular interest is the analysis of this commonly used storage medium with respect to the aroma development of the immersed apple slices. In this context, oxygen availability is one important factor (Beaulieu, 2010; Espino-Díaz et al., 2016). Therefore, the apple slices were also stored in atmospheres with graded oxygen concentrations. The focus of this study lies on the comprehensive evaluation of VOCs emission to indicate the biosynthetic relations affecting aroma-relevant VOCs. For this, apple slices were treated with a commercial ascorbic/citric solution and afterwards stored in air, in three different modified atmospheres (O_2: 10, 5 and approx. 0 %; CO_2: 10, 15 and 20 %) or in sugar syrup (20 %) for up to 10 d. The physiological and microbial quality parameters of samples were analysed on days 3, 6 and 10 of storage. Understanding the respective physiological changes in VOC synthesis will facilitate the extension of the shelf life of fresh-cut fruits.

2. Materials and methods

2.1 Material

Fresh mature 'Braeburn' apples (*Malus domestica* Borkh.) were obtained from a commercial fresh-cut salad producer, transported to the Department of Horticultural Engineering (Leibniz Institute for Agricultural Engineering and Bioeconomy, Potsdam, Germany) and stored at 4 °C and 95 % relative humidity for 1 d until experiments. Apples of uniform size and without any defects were selected for further investigation. The initial mean fresh mass (n = 20) and dry matter content were 147.6 ± 11.2 g and 179 ± 14 g kg^{-1}, respectively.

2.2 Fresh-cut preparation, storage and sampling

All apples were processed under hygienic conditions in a cooling room at 4 °C. The apples were halved equatorially and each half-segment were cut into 16 slices using a Parti apple cutter/corer (Gefu GmbH, Eslohe, Germany). Closely following the processes used in practice, all apple slices were immersed for 5 min in ascorbic/citric solution (40 g ascorbic acid and 20 g citric acid in 1 L deionized water) as usual anti-browning treatment and then divided into 5 batches. Slices of each batch were filled in commercial 840 mL plastic pails or 2.9 L metal steel cuvettes according to five different storage conditions applied (**Table 4.1**). These conditions include storage in normal air (AIR), in three different modified atmospheres (O_2/CO_2: MA1 = 10 %/10 %; MA2 = 5 %/15 %; MA3 = 0 %/20 %) or in sugar syrup (SY, 20 %). Each pail or steel cuvette contained 40 (approx. 228 g) or 120 (approx. 684 g) apple slices, respectively. Pails of the air-stored apples (AIR) were loosely covered with a plastic lid, while that of syrup-immersed apples (SY) were filled up with 450 mL sugar syrup and tightly closed. The commercial sugar syrup (200 g L^{-1} invert sugar syrup, 72.7 %; Hanseatische Zuckerraffinerie GmbH & Co. KG, Hamburg, Germany) contained 10 g L^{-1} OBSTSERVAL HC-2 browning inhibitor (Konserval, Pharmacon Lebensmittelzusätze GmbH, Trittau, Germany), consisting of ascorbic acid, sodium ascorbate and citric acid. All steel cuvettes were airtight closed with acrylic lid and flushed with modified atmospheres (MA, i.e., MA1 to MA3), created with a gas mixer (Mätti Mess- und Regeltechnik, Unna, Germany). All containers were stored at 4 °C and samples were taken on days 3, 6 and 10. For this, three pails (replicates) from each of the AIR and the SY treatments were opened. For MA1 to MA3, the O_2 and CO_2 concentrations of all cuvettes were measured on each sampling day using a CheckMate 3 gas analyser (Dansensor, Ringsted, Denmark) before opening. After opening the cuvettes, 40 apple slices were removed and then the cuvettes closed and flushed again to regain the specified conditions.

Table 4.1: Treatments of fresh-cut apple slices, stored in different containers under ambient air (AIR), modified (MA) and in sugar syrup (SY) at 4 °C for up to 10 d. Given are also the total number of container used per treatment, the Quantity of apple slices per container and the concentrations of O_2 and CO_2 used.

Variant	Container	Quantity/ containers	Number/ slices	O_2	CO_2
AIR	Plastic pail	12	40	20 %	0 %
MA1	Metal steel cuvette	3	120	10 %	10 %
MA2	Metal steel cuvette	3	120	5 %	15 %
MA3	Metal steel cuvette	3	120	0 %	20 %
SY	Plastic pail	12	40	immersed in syrup	

2.3 Sampling and analyses of volatile organic compounds (VOCs)

For each batch and sampling day, VOCs and ethylene were measured in triplicate at 4 °C. For this, a 1 L canning glass jar equipped with a septum was filled with 20 apple slices and hermetically closed to allow the accumulation of VOCs and ethylene. After 1 h, VOCs were extracted for 10 min with 50/30 µm divinylbenzene/carboxen/polydimethylsiloxane (DVB/CAR/PDMS) SPME fibre (stableFlex/SS, Supelco, Bellefonte, USA). Afterwards, 4.5 mL air was taken using 10 mL Omnifix® sample syringe (B. Braun Melsungen AG, Melsungen, Germany) for measurements of ethylene concentration. The SPME fibre was introduced for 30 min into the injector kept at 250 °C of an GCMS-QP2010 SE gas chromatograph – mass spectrometer (GC–MS; Shimadzu Europa GmbH, Duisburg, Germany) equipped with a SPME liner (0.75×5.0×95 mm for Shimadzu GCs; Restek, Bellefonte, PA, USA) and a DB-WAX (Agilent Technologies, Palo Alto, CA, USA) column (30 m×0.25 mm×0.25 µm). VOCs analyses were performed at a constant helium flow of 0.8 mL min^{-1} and a temperature program of 35 °C (hold 1 min), ramp to 110 °C at 5 °C min^{-1} and ramp to 230 °C at 10 °C min^{-1} (hold 3 min). Data were recorded in MS full scan mode (35–250 m/z) at 70 eV. Integration based on quantifier and qualifier ions using MSD Chem-Station E.02.01.1177 (Agilent Technologies), where specific VOCs identification was performed with NIST v2.0f library (NIST, USA) and retention indices (RI) according to van Den Dool and Kratz (1963) comparing with online NIST Chemistry WebBook (**Table 4.2**). For RI calculation, 5 µL of alkane standard (C7 - C30; 1000 µg mL^{-1}; Supelco) was injected into a 3.9 L glass jar and analysed analogy to the apple samples (1 d equilibrium time). Presented semi-quantitative VOC concentrations were calculated based on 2-methylbutyl 3-methylbutanoate (CAS-number: 2245-77-4) as external standard and expressed in nL L^{-1}. Therefore, the total integrated peak areas of the VOCs were extrapolated using a standard curve of 2-methylbutyl 3-methylbutanoate with known concentration (5–1000 nL L^{-1}) to get the relative concentrations.

Table 4.2: All VOCs identified in the presented experiments. Data includes the CAS registry numbers, the calculated retention indices (RI), the retention indices reported in literature ($RI_{Literature}$), and quantifier and qualifier ions used for peak integration.

Chemical Group	VOCs	CAS-Number	RI	$RI_{literature}$	Quantifier (Qualifier) Ions (m z^{-1})
esters (25)	Ethyl acetate	141-78-6	894	863-908	43 (45-61)
	Ethyl propionate	105-37-3	957	939-976	57 (74-75)
	Ethyl 2-methylpropanoate	97-62-1	966	957-969	43 (41-71)
	Propyl acetate	109-60-4	975	952-996	43 (61-73)
	Methyl butyrate	623-42-7	986	969-993	74 (43-71)
	Methyl 2-methylbutyrate	868-57-5	1009		57 (41-88)
	Isobutyl acetate	110-19-0	1013	1000-1031	43 (56-73)
	Ethyl butyrate	105-54-4	1034	1000-1073	71 (43-88)
	Ethyl 2-methylbutanoate	7452-79-1	1049	1022-1073	57 (41-102)
	Butyl acetate	123-86-4	1069	1049-1105	43 (56-73)
	Diethyl carbonate	105-58-8	1102		45 (63-91)
	2-methylbutyl acetate	624-41-9	1116	1111-1125	43 (55-70)
	Ethyl valerate	539-82-2	1129	1131-1139	88 (57-85)
	Butyl propionate	590-01-2	1134	1132-1138	57 (56-75)
	Ethyl 2-butenoate	10544-63-5	1154	1158-1158	69 (41-99)
	Pentyl acetate	628-63-7	1165	1175-1181	43 (55-70)
	Methyl hexanoate	106-70-7	1178	1176-1189	74 (43-99)
	Butyl butyrate	109-21-7	1208	1196-1245	71 (43-89)
	Ethyl hexanoate	123-66-0	1222	1212-1267	88 (43-99)
	Hexyl acetate	142-92-7	1259	1251-1311	43 (56-61)
	5-Hexen-1-yl acetate	5048-26-0	1309		43 (54-67)
	(2Z)-2-Hexen-1-yl acetate	106-30-9	1317	1319-1319	43 (67-82)
	Hexyl propionate	2445-76-3	1321	1317-1332	57 (56-75)
	Hexyl butyrate	2639-63-6	1400	1393-1410	71 (43-89)
	Hexyl hexanoate	6378-65-0	1610	1596-1599	43 (56-117)
ketones	2-Butanone	78-93-3	906	875-926	43 (57-72)
aldehydes	Hexanal	66-25-1	1076	1048-1120	44 (41-56)
alcohols (4)	Ethanol	64-17-5	941	900-955	45 (43-46)
	2-Methyl-1-propanol	78-83-1	1095	1092-1114	43 (41-42)
	2-Methyl-1-butanol	1565-80-6	1200		57 (41-56)
	2-(2-Ethoxyethoxy)ethanol	111-90-0	1628	1615-1619	45 (59-72)
benzenic derivatives	Estragole	140-67-0	1669	1624-1661	148 (117-147)
terpenes (2)	D-Limonene	5989-27-5	1187	1176-1238	68 (67-93)
	α-Farnesene	502-61-4	1748	1720-1764	93 (41-69)

2.4 Determination of ethylene release

Each replication on each sampling day was measured with three repetitions. For sampling preparation, 10 mL air was removed twice from a closed 20 mL GC vial with a 10 mL-Omnifix® sample syringe (B. Braun Melsungen AG), air (4.5 mL) was taken from a canning glass jar containing apple samples (c.f. 2.3) and injected in the evacuated vial. The ethylene concentration in the vial was analysed with the ETD-300 photoacoustic ethylene detector (Sensor sense, Nijmegen, The Netherlands), operating in the sample mode at a continuous airflow of 0.04 L min^{-1}, time resolution of 5 s and detection limit of 0.3 nL L^{-1}. Ethylene concentrations were calculated using a 10 μL L^{-1} -ethylene standard.

2.5 Measurement of CO_2 release

On each sampling day, apple slices (n = 24) from each replication were used to analyse the respiration rates (RR) by measuring the CO_2 release (mg kg^{-1} h^{-1}) with a custom-made gas exchange system (for details see Rux et al., 2017) at 4 °C (storage temperature) for 4 h. For this, apple slices were transferred to a clean plastic pail and placed in one of nine acrylic glass cuvettes (volume 8.2 L). In each cuvette, CO_2 concentrations were measured with GMP222 CO_2 sensors (Vasalia, Helsinki, Finland) and data recorded by a NetDAQ 2645A data logger (Fluke Deutschland GmbH, Glottertal, Germany).

2.6 Other relevant quality parameters

Different quality parameters were determined on 10 apple slices for each sample replication and each sampling day. Tissue colour parameters (CIE L*, a*, and b*) were measured using a CM-2600d spectrophotometer (Konica Minolta Sensing Inc., Tokyo, Japan) and the browning Index, BI (Maskan, 2001), calculated as

$$BI = \frac{[100(x - 0.31)]}{0.17}, \text{ where } \qquad x = \frac{(a + 1.75L)}{(5.645L + a - 3.012b)}$$

Tissue strength was measured with a SMS XT Plus texture analyser (Stable Micro Systems, Godalming, UK), fitted with a SMS-P/4 cylindrical probe, as the maximum compression force (N) at 8 mm indentation. Afterwards, the apple slices were juiced with a garlic press for determination of soluble solid content (SSC), the total titratable acidity (TA) and the vitamin C content (Vit-C). SSC (% Brix) was measured with a DR301-95 electronic refractometer (A.KRÜSS Optronic GmbH, Hamburg, Germany), TA (mg kg^{-1} citric acid) was determined using an automated T50M Titrator (with Rondo 20 sample changer, Mettler Toledo, Gießen, Germany) by titration with 0.1 mol L^{-1} NaOH to pH 8.2, while Vit-C (mg L^{-1}) was obtained with the Reflectoquant® test kit (Merck, Darmstadt, Germany).

2.7 Microbial analysis

Microbial loads were analysed by the total plate count method as described in detail by Rux et al. (2017) and results were expressed as colony-forming unit per gram fresh mass (CFU g^{-1}). For analysis, six apple slices of each replication per sample day were used and evaluated in duplicate (n = 2). Total aerobic mesophilic bacteria (TAMB) were incubated on plate count agar (Carl Roth GmbH & Co. KG, Karlsruhe, Germany) at 30 °C for 3 d. Yeasts and moulds were incubated on rose bengal chloramphenicol agar (Carl Roth GmbH & Co. KG) at 25 °C for 7 d.

2.8 Statistical analysis

For each treatment and sampling day, three replicates ($n = 3$; 40 apple slices each) were analysed. To determine the microbial loads, six apple slices were homogenized in duplicate and results were averaged before the total means were calculated. In addition, 20 apple slices were used to evaluate VOCs and ethylene emissions/release, while 24 slices were used to determine respiration rate, with three replication per each treatment ($n = 3$). Initially, colour and tissue strength were measured on 10 individual apple slices (per each repetition), which were then juiced to determine SSC, TA and Vit-C and results averaged ($n = 3$).

WinSTAT (R. Fitch Software, Staufen, Germany) was used for statistical analyses (ANOVA). All results were presented as means ± standard deviation (SD) and the significance of differences between means calculated with Duncan's multiple range test ($p < 0.05$). Principal components analysis (PCA) was applied to evaluate the relationship between storage condition, storage time and VOCs emission, retaining the maximum amount of variance (Latentix Ver. 2.00; Latent5, Copenhagen, Denmark). For PCA the normalized areas of VOCs measured by SPME/GC-MS were used.

3. Results

3.1 Atmospheric composition

Mean O_2 concentrations of modified atmospheres, measured on each sampling day for apple slices of each treatment, were 9.9 ± 0.1 % (MA1), 5.2 ± 0.2 % (MA2) and 0.3 ± 0.1 % (MA3); the corresponding average CO_2 concentrations were 10.1 ± 0.1 %, 14.9 ± 0.2 % and 20.3 ± 0.3 % for MA1, MA2 and MA3, respectively.

3.2 Volatile organic compounds

In the current study, 34 different VOCs were emitted by 'Braeburn' apples after cutting and during storage under the various conditions (**Table 4.2**). These VOCs mainly include esters (25) and, to a lesser extent, alcohols (4), terpenes (2), ketones (1), aldehydes (1) and benzenic derivatives (1). The quantities of all VOCs were significantly affected by cutting, storage conditions and/or time (**Table 4.3**). In intact apples, VOCs with the highest proportion were ethyl butyrate (17 %), hexyl acetate (16.2 %), 2-methylbutyl acetate (14.6 %) and butyl acetate (11.9 %).

Emission of most VOCs (21) pronouncedly increased after cutting, but their contents often declined again within 4 h of storage. These compounds include most of the esters, two alcohols, hexanal and both terpenes.

Despite the decrease during the initial 4 h, contents of many VOCs, including all eight ethyl esters, as well as methyl hexanoate, ethanol and 2-methyl-1-butanol, 2-butanone (< 10 % oxygen), estragole and both terpenes, significantly increased after the third day of storage. Furthermore, the emission of the majority of VOCs increased during storage. At the end of storage, the total VOCs content, released from air or MA-stored samples, was 5.5 - 6.7 times higher than that emitted by intact apples (**Table 4.3, 1st position**).

For a total of 17 VOCs, contents did not significantly differ between samples stored in air or under MA conditions and no trends of changes were apparent. However, storage at oxygen concentrations clearly below 5 % (MA3 & SY) led to higher emission of ethyl 2-butanoate, methyl hexanoate and ethyl hexanoate, 2-methyl-1-propanol, 2-methyl-1-butanol, hexanal, estragole and D-limonene. Furthermore, MA storage of samples significantly increased the headspace contents of diethyl carbonate, ethyl propionate and ethyl butyrate, but without any apparent effect of the oxygen concentration. On the other hand, VOCs emission was reduced for butyl acetate, isobutyl acetate, 2-methylbutyl-acetate and pentyl acetate, as well as ethyl 2-methylpropanoate under these conditions. The ethanol content significantly increased during anaerobic storage compared to other MA conditions, but not compared to air storage.

Storage of apple slices in syrup let the emission of almost all VOCs continuously decline. On storage day 10, the respective VOCs headspace contents were lower than or similar to those of intact apples. In contrast, the headspace contents of most alcohols increased within 6 d of storage. Differences in the development of VOCs headspace contents in samples stored under atmospheric conditions and those immersed in sugar syrup were prominent. Storage in syrup generally resulted in distinctly lower VOCs emission of samples than storage in air or MA.

On day 10, total VOCs headspace content of syrup-stored samples was 40 % lower than that of intact apples and accounted for only 8.9 – 11 % of samples stored in MA; only contents of ethanol and D-limonene were significantly higher for the syrup stored slices on day 10.

Table 4.3: Semi-quantitative VOCs concentrations, emitted by intact apples and fresh-cut apple slices directly (cut 0h), 4 h after cutting (cut 4h) and after storage in different atmospheric conditions (AIR: ambient air; MA1: 10 %/10 % O_2/CO_2; MA2: 5 %/15 % O_2/CO_2; MA3: 0 %/20 % O_2/CO_2) or sugar syrup (SY at 4 °C) for up to 10 d. Concentrations resulted from the accumulation of VOCs emitted by 20 apple slices in a hermetically closed 1 L glass during 1 h at 4 °C. Given are means (n = 3). Different letters indicate significant differences between means ($p < 0.05$). The different colours may help to indicate increasing (green) or decreasing (red) emissions of VOCs compared to intact apples or (yellow).

VOC	Time	Initial	Time	AIR	MA1	MA2	MA3	SY
Cumulative VOC	intact	**246 a**	3 d	**1638 g-h**	**1050 c**	**1328 e-f**	**1283 d-e**	**348 a**
concentration	cut 0h	**1846 i**	6 d	**1114 c-d**	**1505 f-h**	**1535 f-h**	**1271 d-e**	**225 a**
	cut 4h	**887 b**	10 d	**1344 e-g**	**1662 h**	**1586 g-h**	**1488 f-h**	**147 a**
Diethyl carbonate	intact	**nd**	3 d	**0.03 a**	**0.67 a-b**	**1.15 b**	**1.09 b**	**0.10 a**
es1	cut 0h	**0.06 a**	6 d	**0.08 a**	**3.05 e**	**2.67 d-e**	**2.31 c-d**	**0.04 a**
	cut 4h	**nd**	10 d	**0.03 a**	**2.70 d-e**	**2.81 d-e**	**1.92 c**	**nd**
Ethyl acetate	intact	**15.9 a**	3 d	**239 b**	**258 b-c**	**296 d-e**	**320 e**	**44.8 a**
es2	cut 0h	**47.9 b**	6 d	**280 c-d**	**383 f**	**390 f**	**381 f**	**36.3 a**
	cut 4h	**29.9 a**	10 d	**380 f**	**414 g**	**423 g**	**434 g**	**21.4 a**
Propyl acetate	intact	**3.06 a-b**	3 d	**6.29 d-e**	**4.95 b-e**	**7.01 e**	**5.99 c-e**	**3.87 a-d**
es3	cut 0h	**24.8 g**	6 d	**5.08 b-e**	**4.41 a-e**	**5.55 b-e**	**3.38 a-c**	**2.18 a**
	cut 4h	**14.1 f**	10 d	**6.65 e**	**6.85 e**	**6.98 e**	**5.16 b-e**	**2.02 a**
Butyl acetate*	intact	**29.3 b-d**	3 d	**107 h**	**80.4 f-g**	**100 g-h**	**63.7 e-f**	**52.4 c-e**
es4	cut 0h	**307 j**	6 d	**67.4 e-f**	**53.7 d-e**	**59.6 e-f**	**28.6 b-c**	**25.7 b**
	cut 4h	**147 i**	10 d	**61.2 e-f**	**61.0 e-f**	**52.2 c-e**	**34.2 b-d**	**0.98 a**
Isobutyl acetate	intact	**3.59 a-b**	3 d	**15.1 h**	**11.8 f-h**	**14.1 g-h**	**9.11 d-f**	**10.2 d-f**
es5	cut 0h	**52.7 j**	6 d	**10.6 e-g**	**6.74 b-e**	**7.43 b-e**	**3.47 a-b**	**6.51 b-d**
	cut 4h	**25.2 i**	10 d	**10.4 d-g**	**9.63 d-f**	**7.87 c-e**	**4.94 a-c**	**1.56 a**
2-methylbutyl acetate*	intact	**35.9 a-b**	3 d	**208 g**	**144 e-f**	**172 f**	**117 d-e**	**69.1 b-c**
es6	cut 0h	**621 i**	6 d	**114 d-e**	**105 d**	**115 d-e**	**53.0 b**	**40.0 a-b**
	cut 4h	**277 h**	10 d	**111 d-e**	**116 d-e**	**101 c-d**	**69.3 b-c**	**6.88 a**
Pentyl acetate*	intact	**3.59 a-c**	3 d	**19.6 g**	**12.3 e-f**	**15.2 f**	**10.4 d-f**	**3.41 a-c**
es7	cut 0h	**43.1 h**	6 d	**13.5 e-f**	**12.2 e-f**	**12.5 e-f**	**6.17 b-d**	**1.58 a-b**
	cut 4h	**19.9 g**	10 d	**13.1 e-f**	**13.7 f**	**11.6 e-f**	**8.35 c-e**	**nd**
Hexyl acetate*	intact	**40.0 a-c**	3 d	**202 g**	**117 d-e**	**151 e-g**	**112 d-e**	**15.4 a-b**
es8	cut 0h	**374 h**	6 d	**153 e-g**	**129 d-f**	**122 d-f**	**75.8 b-d**	**7.16 a**
	cut 4h	**184 f-g**	10 d	**139 d-f**	**155 e-g**	**126 d-f**	**92.8 c-e**	**1.09 a**
5-Hexen-1-yl-acetate	intact	**0.70 a-b**	3 d	**3.69 d-e**	**1.97 a-d**	**2.72 b-e**	**1.73 a-d**	**0.40 a**
es9	cut 0h	**7.34 f**	6 d	**2.85 c-e**	**2.06 a-d**	**2.01 a-d**	**1.06 a-c**	**0.17 a**
	cut 4h	**4.30 e**	10 d	**2.64 b-e**	**2.62 b-e**	**1.94 a-d**	**1.35 a-c**	**0.18 a**
(2Z)-2-Hexen-1-yl-acetate	intact	**0.30 a**	3 d	**0.06 a**	**nd**	**nd**	**1.18 a**	**0.31 a**
es10	cut 0h	**22.3 c**	6 d	**nd**	**nd**	**nd**	**nd**	**nd**
	cut 4h	**10.3 b**	10 d	**nd**	**nd**	**nd**	**nd**	**0.08 a**
Ethyl propionate	intact	**4.63 a-b**	3 d	**20.3 e-f**	**13.7 c**	**19.2 d-e**	**16.7 c-e**	**4.38 a-b**
es11	cut 0h	**12.1 c**	6 d	**14.6 c-d**	**24.5 f-g**	**28.4 g-h**	**20.6 e-f**	**2.86 a-b**
	cut 4h	**6.83 b**	10 d	**18.8 d-e**	**30.3 h**	**34.4 i**	**28.6 g-h**	**0.87 a**
Ethyl 2-methylpropanoate	intact	**0.69 a**	3 d	**14.2 f**	**5.92 b**	**5.43 b**	**4.83 b**	**1.47 a**
es12	cut 0h	**2.34 a**	6 d	**12.1 d-e**	**12.8 e-f**	**9.83 c**	**6.95 b**	**1.12 a**
	cut 4h	**1.06 a**	10 d	**18.1 h**	**24.7 i**	**16.2 g**	**10.5 c-d**	**1.57 a**
Butyl propionate*	intact	**2.59 b**	3 d	**1.73 a-b**	**1.24 a-b**	**2.57 b**	**1.63 a-b**	**1.81 a-b**
es13	cut 0h	**15.0 d**	6 d	**0.85 a**	**1.08 a-b**	**1.27 a-b**	**1.41 a-b**	**1.06 a-b**
	cut 4h	**6.36 c**	10 d	**0.85 a**	**1.10 a-b**	**0.97 a-b**	**1.53 a-b**	**0.90 a**
Hexyl propionate*	intact	**2.76 a**	3 d	**nd**	**nd**	**nd**	**nd**	**nd**
es14	cut 0h	**2.72 a**	6 d	**nd**	**nd**	**nd**	**nd**	**nd**
	cut 4h	**1.15 b**	10 d	**nd**	**nd**	**nd**	**nd**	**nd**
Methyl butyrate	intact	**2.53 a-c**	3 d	**5.24 b-d**	**2.91 a-c**	**5.06 b-d**	**6.01 c-e**	**3.22 a-d**
es15	cut 0h	**11.5 f**	6 d	**2.83 a-c**	**4.46 b-d**	**5.89 c-e**	**4.60 b-d**	**1.98 a-b**
	cut 4h	**6.07 c-e**	10 d	**4.61 b-d**	**8.90 e-f**	**9.74 f**	**6.55 d-e**	**0.24 a**

Table 4.3. *Cont.*

VOC	Time	Initial	Time	AIR	MA1	MA2	MA3	SY
Methyl 2-methylbutyrate*	intact	1.19 a-b	3 d	5.60 c-d	2.03 a-c	3.90 a-c	5.31 b-d	1.46 a-c
es16	cut 0h	8.04 d	6 d	2.38 a-c	3.52 a-c	4.17 a-c	3.34 a-c	1.09 a-b
	cut 4h	3.86 a-c	10 d	3.90 a-c	5.15 a-d	4.60 a-d	3.93 a-c	1.00 a
Ethyl butyrate*	intact	41.8 a	3 d	339 e-f	148 b	221 c-d	222 c-d	45.6 a
es17	cut 0h	152 b	6 d	187 b-c	339 e-f	378 f-g	287 d-e	30.8 a
	cut 4h	71.8 a	10 d	263 d	405 f-g	416 g	345 e-f	16.4 a
Ethyl 2-butenoate	intact	nd	3 d	1.43 a-c	0.78 a-b	1.60 a-c	4.10 d	0.29 a
es18	cut 0h	nd	6 d	1.65 a-c	2.75 b-d	4.19 d	8.90 f	0.23 a
	cut 4h	nd	10 d	3.01 c-d	4.51 d	7.06 e	21.3 g	0.24 a
Ethyl 2-methylbutanoate*	intact	12.4 a	3 d	254 i	85.2 c-d	104 d-e	115 d-e	23.3 a-b
es19	cut 0h	63.2 b-c	6 d	124 d-f	167 g-h	157 f-h	134 e-g	15.9 a
	cut 4h	28.7 a-b	10 d	160 f-h	191 h	156 f-h	169 g-h	16.9 a
Butyl butyrate*	intact	6.85 b	3 d	4.74 c	2.18 d	3.06 d	2.76 d	2.48 d
es20	cut 0h	14.0 a	6 d	2.03 d	2.27 d	2.88 d	1.55 d-e	1.84 d-e
	cut 4h	6.72 b	10 d	1.45 d-e	3.05 d	2.54 d	1.76 d-e	nd
Hexyl butyrate*	intact	8.67 a	3 d	3.36 c-d	2.79 c-e	3.45 c-d	2.54 c-e	2.80 c-e
es21	cut 0h	6.66 b	6 d	1.38 d-e	2.03 c-e	2.89 c-e	1.83 c-e	2.15 c-e
	cut 4h	3.69 c	10 d	1.67 c-e	2.98 c-e	2.84 c-e	2.49 c-e	0.79 e
Ethyl valerate	intact	0.03 a	3 d	3.15 g-h	1.64 c-d	2.68 e-f	2.52 e	0.31 a-b
es22	cut 0h	0.69 b	6 d	1.28 c	3.18 g-h	3.07 f-g	2.82 e-g	0.25 a-b
	cut 4h	0.35 a-b	10 d	2.01 d	3.51 h	3.49 h	2.93 e-g	nd
Methyl hexanoate	intact	0.53 a	3 d	3.84 b-c	4.84 c	6.99 d-e	12.8 f	1.05 a
es23	cut 0h	2.03 a-b	6 d	1.89 a-b	8.33 e	5.63 c-d	7.45 d-e	0.64 a
	cut 4h	1.24 a	10 d	2.47 a-b	8.06 e	7.16 d-e	8.94 e	0.88 a
Ethyl hexanoate	intact	4.06 a	3 d	107 d-e	85.8 c-d	116 d-f	149 g	20.5 a
es24	cut 0h	20.0 a	6 d	52.9 b	152 g	141 f-g	138 e-g	8.52 a
	cut 4h	13.5 a	10 d	72.7 b-c	132 e-g	128 e-g	145 f-g	2.38 a
Hexyl hexanoate*	intact	3.85 a	3 d	3.24 a-b	3.30 a-b	3.59 a	3.15 a-c	3.02 a-d
es25	cut 0h	2.37 b-f	6 d	1.96 d-f	2.22 b-f	2.83 a-e	1.75 e-f	2.10 c-f
	cut 4h	1.87 e-f	10 d	1.23 f	2.08 c-f	2.87 a-e	2.03 d-f	1.41 f
Ethanol	intact	7.27 a	3 d	31.6 b	32.0 b	34.8 b	51.0 c-d	10.1 a
al1	cut 0h	2.19 a	6 d	34.3 b	55.3 c-d	42.3 b-c	61.3 d	11.7 a
	cut 4h	4.77 a	10 d	45.8 b-c	36.5 b	36.4 b	54.1 c-d	51.1 c-d
2-Methyl-1-propanol	intact	0.70 a	3 d	1.23 b-c	1.10 b-c	1.16 b-c	1.32 c	0.65 a
al2	cut 0h	1.66 d	6 d	1.25 b-c	1.00 b	1.02 b	1.25 b-c	0.69 a
	cut 4h	1.56 d	10 d	1.22 b-c	1.04 b	1.11 b-c	1.56 d	1.06 b
2-Methyl-1-butanol	intact	3.68 a-b	3 d	9.96 g-h	5.79 c-d	7.69 e-f	11.6 i	2.97 a
al3	cut 0h	9.46 g-h	6 d	5.79 c-d	7.85 e-f	8.68 f-g	10.7 h-i	2.97 a
	cut 4h	7.52 e-f	10 d	4.95 b-d	5.10 b-d	6.6 d-e	10.8 h-i	4.56 a-c
2-(2-Ethoxyethoxy)ethanol	intact	0.57 b-c	3 d	0.23 a	0.71 c-d	0.50 b	0.27 a	0.96 e
al4	cut 0h	0.56 b-c	6 d	0.55 b-c	0.46 b	0.55 b-c	0.55 b-c	0.80 d
	cut 4h	0.58 b-c	10 d	0.51 b	0.48 b	0.62 b-c	0.50 b	0.60 b-c
2-Butanone	intact	0.41 a-b	3 d	0.84 a-c	1.47 a-d	1.96 c-d	6.52 e	0.66 a-c
k1	cut 0h	0.08 a	6 d	0.78 a-c	1.24 a-d	1.18 a-d	2.15 d	0.42 a-b
	cut 4h	0.13 a	10 d	0.46 a-b	0.65 a-c	0.81 a-c	1.65 b-d	0.57 a-b
Hexanal*	intact	0.55 a-c	3 d	0.26 a-b	0.11 a	0.76 b-c	0.69 a-c	1.99 d
ad1	cut 0h	4.45 e	6 d	0.29 a-b	0.26 a-b	0.35 a-b	0.66 a-c	2.11 d
	cut 4h	0.56 a-c	10 d	0.39 a-b	0.54 a-c	0.60 a-c	1.01 c	0.44 a-c
Estragole	intact	0.66 b-d	3 d	1.24 f	0.59 b-c	0.79 c-e	0.87 d-e	0.67 b-d
bd1	cut 0h	0.75 c-d	6 d	0.63 b-d	0.79 c-e	0.80 c-e	0.82 c-e	0.57 b-c
	cut 4h	0.31 a	10 d	0.62 b-d	0.78 c-e	0.85 d-e	0.99 e	0.45 a-b
D-Limonene	intact	0.64 a-c	3 d	2.53 g	1.63 d-e	1.77 d-f	2.39 g	2.36 f-g
tp1	cut 0h	1.23 b-d	6 d	0.88 a-c	1.62 d-e	1.61 d-e	2.47 g	1.66 d-e
	cut 4h	0.74 a-c	10 d	0.50 a	0.56 a-b	0.75 a-c	1.92 e-g	1.31 c-e
α-Farnesene	intact	7.03 a	3 d	22.2 g	16.3 d-e	20.1 f-g	18.6 e-f	15.6 d-e
tp2	cut 0h	13.4 c-d	6 d	15.3 d-e	14.1 c-d	15.4 d-e	15.6 d-e	13.6 c-d
	cut 4h	6.39 a	10 d	11.3 b-c	13.5 c-d	13.0 c-d	14.2 c-d	9.37 a-b

* marked VOCs described as generally important "character impact" compounds in apples (Dixon and Hewett, 2000). nd = not detected/below detection limit; dark green = ≥500 % increase; light green = ≥33 % increase; yellow = ≤33 % increase/decrease; light red = ≥80 % decrease; dark red = ≥500 % decrease.

PCA analyses provided two principal components (PCs), which represented 98.2 % of the total variance of VOCs profiles (PC1 and PC2 represented 65.3 % and 33.0 % of the variation, respectively). The scatter point plot (**Fig. 4.1A**) visualized the effects of storage conditions, storage time and cutting. The cutting-induced alteration of the VOCs profiles can be indicated by the decrease in both PCs. The increase in PC1, combined with the decrease in PC2 can be distinctly associated with the effects of storage time and storage conditions.

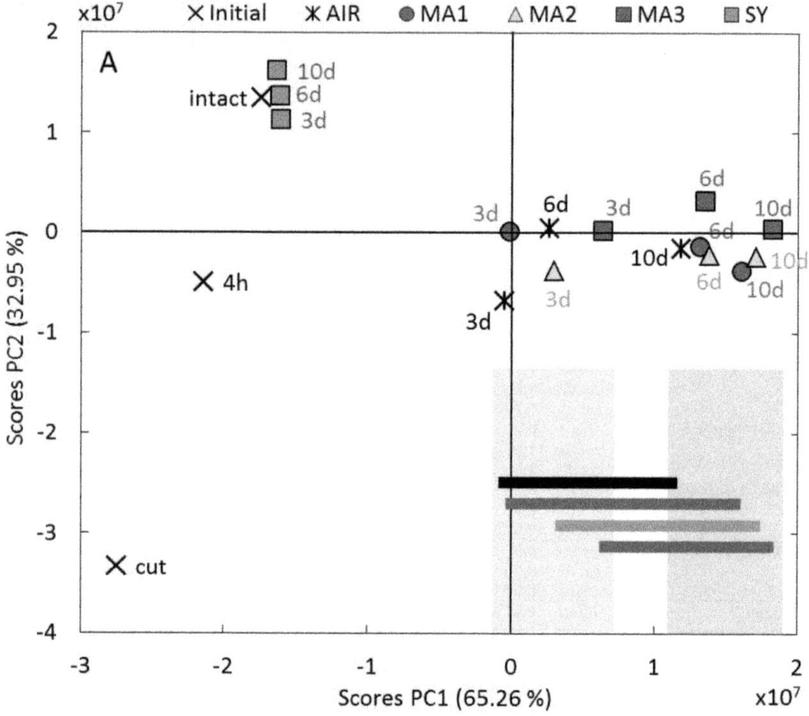

Figure 4.1. (A) PCA scatter point plot of the main sources of variability in VOCs profiles of intact apples and fresh-cut apple slices directly (cut 0h) and 4 h after cutting (cut 4h) and after storage in ambient air (AIR), modified atmospheres (MA1: 10 %/10 O_2/CO_2; MA2: 5 %/15 % O_2/CO_2; MA3: 0 %/20 % O_2/CO_2) or sugar syrup (SY) at 4 °C for up to 10 d. Different scores can be associated with effects of cutting (both PCs scores decreasing) and distinct effects of storage time (increasing PC1 scores) as marked by coloured areas (day 3 = blue vs. day 10 = red) and distinction between the various storage conditions (increasing PC1 and decreasing PC2 scores) marked by coloured bars.

The impact of each single VOC on the total variability in VOCs profiles was visualized in **Fig. 4.1B**. Ethyl acetate (es2) is mainly related to differences between syrup and atmosphere storage, and effects of oxygen availability and of storage time. Similar effects were associated with ethyl butyrate (es17), ethyl 2-methylbutanoate (es19) and hexyl propionate (es24), though to a much smaller extent. Cutting was mainly characterised by the increased emission of 2-methylbutyl acetate (es6), and to a lesser extent by butyl acetate (es4) and hexyl acetate (es8).

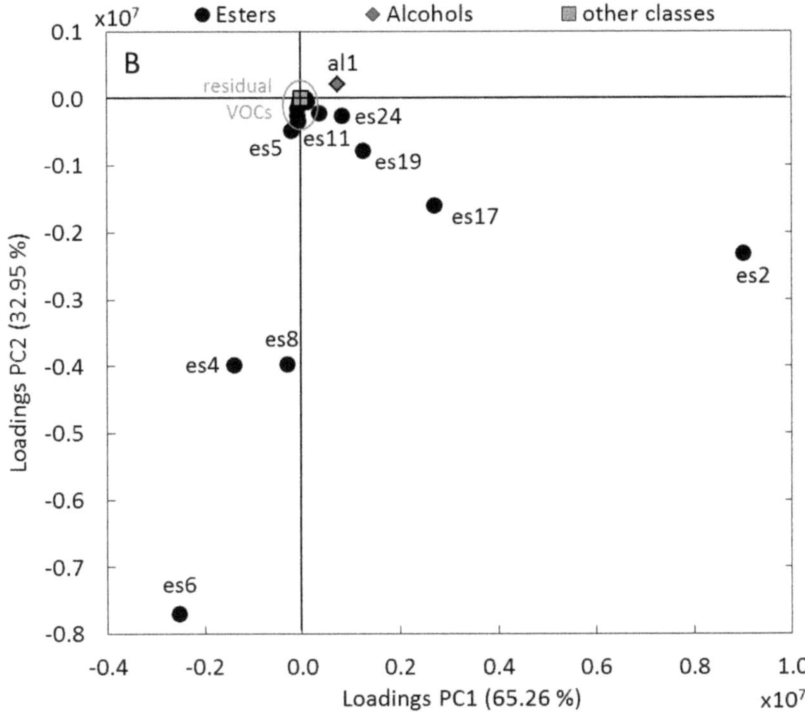

Figure 4.1. (B) Impact of each single VOC on the total variability in VOCs profiles (loadings). VOCs with negative loadings of both PCs can be associated with the effects of cutting; VOCs with positive PC1 and negative PC2 loadings are related to effects of storage time and storage conditions.

3.3 Ethylene release

The mean ethylene release rates of 'Braeburn' apples significantly increased after cutting and within the initial 4 h of storage, but significantly declined below that of intact apple during prolonged storage (**Fig. 4.2**). Compared to other conditions, storage of apple slices under nearly anaerobic conditions (MA3) led to a significant increase in ethylene release as measured after these samples had been moved to normal atmospheric conditions. Highest rates were measured on day 6, when ethylene release was 6 times higher than 4 h after cutting. All other storage conditions (i.e. AIR, MA1, MA2, SY) significantly and continuously decreased the ethylene release of samples after storage, i.e., enhancing the oxygen concentrations during storage may increasingly lower the ethylene release of apple slices.

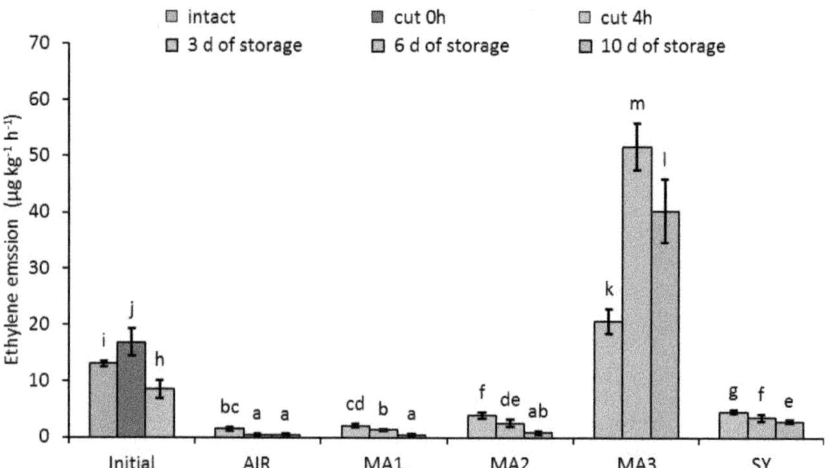

Figure 4.2. Ethylene release of intact apples and fresh-cut apple slices directly (cut 0h), 4 h after cutting (cut 4h) and after storage in ambient air (AIR), modified atmospheres (MA1: 10 %/10 % O_2/CO_2; MA2: 5 %/15 % O_2/CO_2; MA3: 0 %/20 % O_2/CO_2) or sugar syrup (SY) at 4 °C for up to 10 d. Given are means ± standard deviation (n = 3). Different superscripts indicate significant differences between means (Duncan's test, $p < 0.05$).

3.4 Respiration

Mean respiration of intact 'Braeburn' apples increased by 220 % after cutting and remained constant during the initial 4 h of storage (**Fig. 4.3**). Generally, the respiration of apple slices slightly decreased during storage irrespective of storage conditions applied. However, in AIR-samples, respiration drastically increased after 10 d of storage.

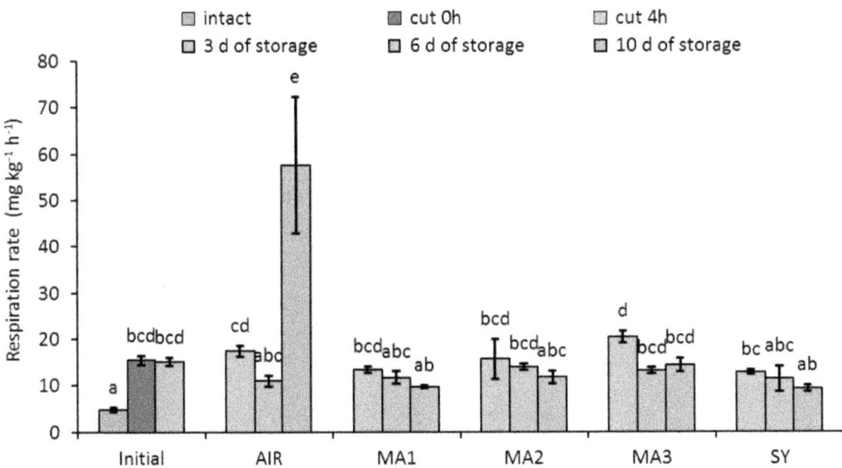

Figure 4.3. Respiration rates of intact apples and fresh-cut apple slices directly (cut 0h), 4 h after cutting (cut 4h) and after storage in ambient air (AIR), modified atmospheres (MA1: 10 %/10 % O_2/CO_2; MA2: 5 %/15 % O_2/CO_2; MA3: 0 %/20 % O_2/CO_2) or sugar syrup (SY) at 4 °C for up to 10 d. Given are means ± standard deviation (n = 3). Different superscripts indicate significant differences between means (Duncan's test, $p < 0.05$).

3.5 Relevant quality parameters

Low atmospheric oxygen concentrations during storage of MA3 (i.e. 0 %) and SY samples prevented tissue browning (BI, **Table 4.4**). In other samples, the browning index, BI, significantly increased with the oxygen concentrations during storage, with browning being largest in air-stored apple slices (i.e. 21 % O_2).

The initial tissue strength of fresh-cut apple slices significantly declined (16 - 27 %) in all samples after 3 d of storage. Only low atmospheric oxygen concentrations let the tissue strength continuously decline during storage (total of 32 % and 55 %, respectively). Neither treatment conditions nor storage time affected the soluble solids content (SSC) of the apple slices.

The total acidity (TA) slightly increased in air and MAP-stored (MA1 to MA3) samples on days 3 and 6. However, these storage conditions did not affect the Vitamin C content of the apple slices, while immersion in sugar syrup drastically increased this parameter.

3.6 Microbial analysis

Generally, microbial loads were very low for all groups of microorganisms in this experiment and no microbes were detectable for initial samples and on day 3 of storage (**Fig. 4.4**). TAMB and yeast counts continuously increased during storage. Low atmospheric oxygen concentrations (≤5 %, MA2 & MA3) significantly reduced the growth of TAMB (**Fig. 4.4A**), whereas 0 % oxygen also led to significantly lower yeast counts on storage day 10 (**Fig. 4.4B**). In contrast, storage in syrup (SY) resulted in significantly higher yeast counts on days 3 and 6. Only under this condition, yeasts were already detectable on storage day 3. On day 10, yeast counts of SY-samples and of those stored in atmospheres containing oxygen concentrations ≥10 % (MA1& AIR) were not different. Mould counts were always below detection limits and only in AIR-samples 2.2 log CFU g^{-1} were determined after 10 d of storage (data not shown).

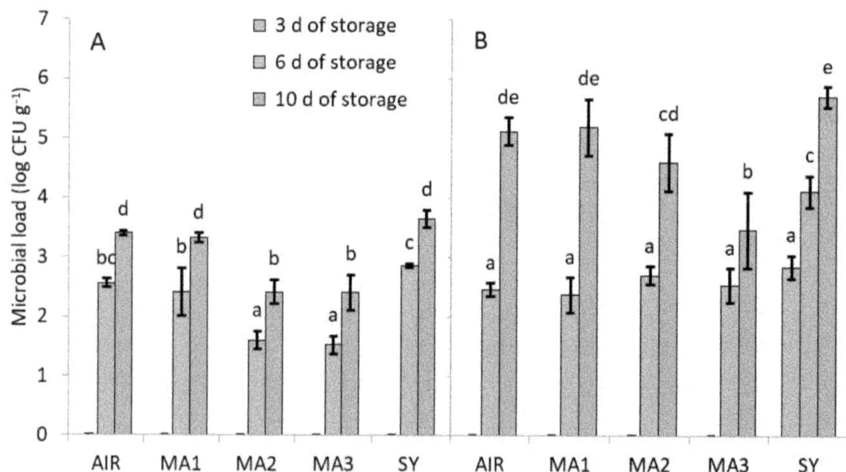

Figure 4.4. Microbial analysis of fresh-cut apple slices stored in air (AIR), modified atmospheres (MA1: 10 %/10% O_2/CO_2; MA2: 5 %/15 % O_2/CO_2; MA3: 0 %/20 % O_2/CO_2) or sugar syrup (SY) at 4 °C for up to 10 d. **(A)** Total aerobic mesophilic bacteria (TAMB) and **(B)** yeast counts. Error bar represents standard deviation of means (n = 3). Different superscripts indicate significant differences between means (Duncan's test, $p < 0.05$).

Table 4.4: Quality parameters (BI: browning index, SSC: total soluble solids, TA: total acidity) of intact apples (initial) and fresh-cut apple slices stored in ambient air (AIR), in modified atmospheres (MA1: 10 %/10% O_2/CO_2; MA2: 5 %/15 % O_2/CO_2; MA3: 0 %/20 % O_2/CO_2) or in sugar syrup (SY) for up to 10 d at 4 °C. Given are means (n = 3) ± standard deviation. Different letters significant differences (Duncan's test, $p < 0.05$) between means within treatments and storage time.

Parameter	Initial	Time (days)	AIR	MA1	MA2	MA3	SY
BI	51.1 ± 9.8 (a)	3	62.3 ± 3.7 (bcd)	61.9 ± 7.7 (bcd)	58.2 ± 6.1 (abc)	56.4 ± 3.3 (abc)	56.2 ± 5.6 (abc)
(rel. units)		6	59.6 ± 2.1 (abcd)	54.9 ± 6.9 (abc)	54.1 ± 3.5 (abc)	52.9 ± 0.7 (abc)	53.8 ± 6.6 (abc)
	8.7 ± 1.0 (a)	10	70.7 ± 8.3 (d)	64.5 ± 11.3 (cd)	63.8 ± 8.8 (cd)	53.5 ± 1.9 (abc)	49.3 ± 3.2 (a)
Tissue strength		3	7.3 ± 0.2 (b)	7.1 ± 0.3 (bc)	6.9 ± 0.3 (bcd)	6.9 ± 0.2 (bcd)	6.3 ± 0.2 (cde)
(N)		6	7.1 ± 0.3 (bc)	6.9 ± 0.2 (bcd)	6.7 ± 0.5 (bcd)	6.2 ± 0.6 (de)	5.0 ± 0.7 (f)
	12.0 ± 0.8 (ab)	10	6.4 ± 0.3 (cde)	7.0 ± 0.1 (bc)	6.9 ± 0.2 (bcd)	5.9 ± 0.5 (e)	3.9 ± 0.2 (g)
SSC		3	11.3 ± 0.9 (ab)	10.0 ± 2.2 (a)	10.7 ± 2.1 (ab)	12.1 ± 0.9 (ab)	11.4 ± 0.5 (ab)
(%Brix)	5.1 ± 0.1 (ab)	6	11.7 ± 0.5 (ab)	12.0 ± 0.4 (ab)	11.4 ± 0.6 (ab)	11.5 ± 1.0 (ab)	11.3 ± 0.3 (ab)
		10	12.5 ± 1.3 (b)	11.7 ± 1.0 (ab)	11.6 ± 1.7 (ab)	11.4 ± 0.6 (ab)	11.1 ± 1.1 (ab)
TA	83.7 ± 34.5 (a)	3	5.6 ± 0.4 (a)	5.5 ± 0.3 (a)	6.9 ± 0.5 (b)	6.8 ± 0.5 (b)	5.4 ± 0.9 (a)
(g kg-1)		6	7.2 ± 1.2 (b)	5.0 ± 0.8 (a)	6.8 ± 0.4 (b)	8.5 ± 1.2 (c)	5.5 ± 0.5 (a)
		10	5.5 ± 0.2 (a)	5.1 ± 0.4 (a)	5.4 ± 0.5 (a)	5.6 ± 0.4 (a)	4.8 ± 0.7 (a)
Vitamin C		3	123 ± 33 (a)	128 ± 19 (a)	141 ± 31 (a)	132 ± 17 (a)	867 ± 256 (b)
(mg L-1)		6	142 ± 29 (a)	127 ± 40 (a)	104 ± 17 (a)	160 ± 30 (a)	1264 ± 50 (c)
		10	138 ± 9 (a)	103 ± 9 (a)	126 ± 11 (a)	124 ± 6 (a)	1161 ± 100 (c)

4. Discussion

4.1 VOCs profiles

The identified esters emitted by intact apples and their high proportion on the VOCs profiles corresponded with previous reports (Paillard, 1990; Mpelasoka and Behboudian, 2002; Baiamonte et al., 2016). In this context, butyl acetate, 2-methylbutyl acetate and hexyl acetate were also identified by others as the most important esters emitted from fresh 'Braeburn' apples (Baiamonte et al., 2016). The measured profiles

also include 13 "character impact" volatile compounds generally described for apples (Dixon and Hewett, 2000).

Extremely low microbial loads of apple slices after processing minimized the microbial impact on VOCs profiles. The low microbial loads were probably due to the low initial contaminations of intact apples and the sanitising effects of the acid treatment (Rahman et al., 2011; Chen et al., 2016).

4.2 Effect of cutting

The strongly increased ester production, especially acetate-esters, and the boost of ethylene release and of respiration induced by the cutting process, can be explained by wound stress responses of fruit resulting in accelerated metabolism (Soliva-Fortuny et al., 2005; Mahajan et al., 2014). In this context, Bett et al. (2001) reported an intensification of the flavour of fresh-cut 'Gala' apple within the first days of storage. According to Schreier (1984), VOCs synthesis is shifting from β-oxidation to lipoxygenase after disruption of fruit tissue, and thereby may provide substrates for ester production (De Pooter et al., 1983). Additionally, the accelerated glycolysis (identified by the enhanced respiration rates) increases the acetyl-CoA availability, which leads to enhanced syntheses of amino-acids, precursors to many VOCs (Beaulieu, 2006). The compound mainly emitted after cutting (i.e. 2-methylbutyl acetate) is primarily produced from amino acids degradation (Rowan et al., 1999). The further decline in ester and ethylene release after cutting reflects other reports that the amplification of the activity of the above mentioned metabolic pathways is only temporary and turns back to normal level within 24 h (Finnegan et al., 2013). However, the rate of decline in acetate ester emissions (except ethyl acetate) was affected by storage conditions (see below).

4.3 Effects of storage in ambient air or under modified atmospheres

Atmospheric storage in either ambient air or under modified atmosphere (MA) led to increased emissions of ethanol and ethyl esters, especially of ethyl acetate, loss of tissue strength, and, depending on oxygen availability, browning and microbial growth. Storage time, modified atmospheres or acid treatment did not affect SSC, TA or Vit-C contents, as also reported elsewhere (Buta et al., 1999; Soliva-Fortuny and Martin-Belloso, 2003a; Rocculi et al., 2004; Cliff et al., 2010; Rux et al., 2019b).

The post-processing acids treatment strongly reduced browning of the fresh-cut apple slices. This was also described by other reports (Soliva-Fortuny and Martin-Belloso, 2003a). Moreover, results obtained indicated that decreasing the steady-state oxygen concentrations from 21 % to 0 % in the storage atmosphere increasingly inhibited the continuous browning of apple slices. Compared to ambient air, reduced oxygen and/or elevated carbon dioxide concentrations may prevent enzymatic browning (Gorny, 2001). This may be attributed to the decreased oxidation of phenolic compounds to the corresponding o-quinones (Gacche et al., 2006). The colour preservative effects of MA were also confirmed by Soliva-Fortuny et al. (2002) and Cortellino et al. (2015).

Reduced microbial growth at very low O_2 concentrations (<5 %) can be directly associated with the inhibiting effects of MA (Farber, 1991; Brackett, 1997). Anyway, the microbial loads at the end of storage were clearly below the standard thresholds defined by The German Association for Hygiene and Microbiology (DGHM, 2011) for TAMB (7 log CFU g^{-1}), yeasts (5 log CFU g^{-1}) and moulds (3 log CFU g^{-1}), respectively.

Reduction of tissue strength, observed for all apple samples irrespective of the treatment at the end of storage can be attributed to changes in physical and mechanical cell wall properties, and tissue structures (Toivonen and Brummel, 2008). These changes may result from the acid treatment. Acidification of tissues is known to induce structural breakdown and, thus, softening (Cocci et al., 2006; Cortellino et al., 2015). Consistently with the presented results, Cocci et al. (2006) indicated negligible effects of MA (5 %/5 % O_2/CO_2) on tissue “firmness”.

In the present study, low O_2 concentrations (<5 %) did not reduce respirational CO_2 release. The very low O_2 concentrations within the fruit tissue (<2 %) did also not restrict ethylene synthesis. These results are in contrast to other studies, which indicated that the above conditions slowed down respiration and led to reduced ethylene release due to strict O_2 requirement of its synthesis (Soliva-Fortuny et al., 2005; Both et al., 2014). These apparently contrasting finding may be attributed to different experimental conditions. After removal from MA storage, respiration and ethylene evolution of the apple slices might have been rapidly adapted to normal air conditions. It was indeed shown that respiration of 'Fuji' apple slices rapidly increased after exposition to ambient air (Gil et al., 1998). The final pronounced increase in respiration of AIR-samples may indicate the initiation of senescence as reported by Rux et al. (2017) for fresh-cut 'Elstar' apples after 6 d of storage.

As stated above, it is obvious from the presented results that the lower the O_2 concentrations was during storage, the higher the ethylene synthesis was after removal to normal air. Earlier reports on the inhibition of ethylene synthesis at very low O_2 concentrations (Soliva-Fortuny et al., 2005) are in accordance with this interpretation. This is also supported by the distinct VOCs emissions under the different storage conditions in this study because ethylene is involved in the synthesis of relevant VOCs (Schaffer et al., 2007). Among others, ethylene directly regulates the synthesis of butyl acetate, 2-methylbutyl acetate, hexyl acetate, ethyl 2-methylpropanoate and ethyl butyrate (Beaulieu, 2006; Obando-Ulloa et al., 2009). These VOCs were significantly reduced at O_2 concentrations close to 0 % (MA1 & SY) and may, thus, reflect a low ethylene synthesis during storage, but not under higher O_2 availability and under AIR conditions (MA2, MA3 & AIR).

The enhanced formation of ethyl esters, as observed in this study, may result from increased contents of ethanol and its esterification by alcohol-acyltransferase (AAT) (Forney et al., 2000). In turn, the increased ethanol emissions in the stored apple slices may be explained by the pronounced O_2 shortage and membrane damage, which could disturb the respiratory metabolism and enhanced the accumulation of pyruvate, the precursor of ethanol (Ke et al., 1994). The odour threshold of ethyl acetate is reported to be low (5-13500 nL L^{-1}; Dixon and Hewett, 2000). Increasing its concentrations create a chemical-like odour associated with anaerobic fruit (Forney et al., 2000). In fact, Mattheis et al. (1991) identified ethyl acetate as the most prevalent ethyl ester that increased in fruit under anaerobic conditions.

Both the increase in ethyl ester and the decrease in acetate ester concentrations were shown here and reported in other studies (Cortellino et al., 2015). The formation of ethyl esters from ethanol consumes high amounts of acetyl-CoA, and thereby decreases concentrations of other acetate esters due to competitive reaction (Kollmannsberger and Berger, 1992; Ke et al., 1994).

4.4 Effect of sugar syrup immersion

In contrast to atmospheric storage, immersion of apple slices in syrup clearly reduced the emission of almost all VOCs, better prevented browning and increased the Vitamin-C content, but enhanced tissue softening and microbial growth. The higher Vitamin-C content directly reflected the uptake of ascorbic acid of apple slices from the syrup. The enhanced acidity may facilitate degradation of pectins, which is most probably responsible for softening (Zhu et al., 2007).

Excellent browning prevention was also shown by Rux et al. (2017) for 'Elstar' apple slices stored in 20 % sugar syrup (4°C; 6 d). This effect was attributed to reduced oxygen availability and the additives contained in the syrup (Nicolas et al., 1994). On the other hand, the enhanced microbial growth, especially for yeasts, may have been supported by the increased availability of water and sugar as a substrate for the microbes. Immersion of apple slices in 20 % sugar syrup was reported to increase TAMB and yeast growth compared to air-storage (Rux et al., 2017). Nevertheless, microbial loads were always below the DGHM (2011) thresholds and can be considered as uncritical in this investigation.

The reduction of VOCs emissions in syrup-stored apple slice resulted in both a pronounced loss of aroma and prevention of off-odour. This could partially be explained by generally reduced VOC emissions due to the enhanced gas diffusion resistance of the liquid (Rux et al., 2017). However, this effect demands the reduction of all VOCs; consequently, it represents only a part of the reductions as the emissions of some VOCs were not reduced, e.g. butyl-propionate or D-Limonene, especially on day 3. In addition, the emissions of ethyl esters were further reduced during storage in sugar syrup, although they increased in AIR- and MA-samples. This reflects metabolic changes, which are independent of O_2 availability.

Moreover, the actual headspace volume may play also an important role. In case of SY-samples, the total "headspace volume" corresponded to only that of the intracellular space. This may lead to significantly higher CO_2 concentrations (>20 %) within the fruit tissue, compared to AIR- and MA-samples, where CO_2 can be released and balanced into the container headspace. Increased CO_2 concentrations in the tissue lowered pH and, therefore may partially inhibit ACC and associated esterification (Ke et al., 1994). VOCs can also not diffuse into the container's headspace and will, therefore, accumulate in the tissue. These increased VOCs concentrations may disturb the equilibrium of the reactions, thus inducing a feedback inhibition of the synthesis of distinct VOCs.

5. Conclusions

This study evaluated quality-relevant physiological and microbiological parameters of fresh-cut 'Braeburn' apples stored either in air, in three different modified atmospheres or in sugar syrup and especially focused on aroma-relevant volatiles. Both fruit cutting and storage of apple slices induced different and clearly distinct physiological responses. Among others, postharvest handling pronouncedly altered the respective VOCs profiles. Storage of apple slices in air or MA led to increased emissions of ethanol and ethyl esters, especially ethyl acetate, to loss of tissue strength, and, depending on O_2 availability, browning and microbial growth. Due to its low aroma threshold, enhanced formation of ethyl acetate may create off-odour and negatively affect the customers' acceptance. In contrast, storing apple slices in syrup clearly reduced the emission of almost all VOCs and better prevented browning, but enhanced tissue softening and microbial growth. VOCs reduction could only be partially explained by enhanced gas diffusion resistance in liquid and indicated metabolic changes, which were independent of O_2 availability. Increased CO_2 or VOCs concentrations in the tissue might have inhibited esterification or might have induced a feedback inhibition of VOCs synthesis, respectively. The reduction of VOCs emissions resulted in both, a pronounced loss of aroma and prevention of off-odour. Therefore, the use of sugar syrup for storage of apple slices might extend shelf life, though at the expense of a lower aroma quality. Storage in sugar syrup might be preferable for apple cultivars that are especially prone to fast browning, but easily retain firmness in shelf life.

5 Effects of pre-processing short-term hot-water treatments on quality and shelf life of fresh-cut apple slices

Article III

Guido Rux [1,2], Efecan Efe [1,2], Christian Ulrichs [2], Susanne Huyskens-Keil [2], Karin Hassenberg [1] and Werner B. Herppich [1,*]

[1] Department of Horticultural Engineering, Leibniz Institute for Agricultural Engineering and Bioeconomy (ATB), Max-Eyth-Allee 100, 14469, Potsdam, Germany

[2] Thaer-Institute of Agricultural and Horticultural Sciences, Division Urban Plant Ecophysiology, Section Quality Dynamics/Postharvest Physiology, Humboldt-Universität zu Berlin, Lentzeallee 55/57, 14195 Berlin, Germany

* Correspondence: wherppich@atb-potsdam.de; Tel.: +49-331-5599-620

Received: 18 October 2019; Accepted: 25 November 2019; Published: 6 December 2019

Abstract: Processing, especially cutting, reduces the shelf life of fruits. In practice, fresh-cut fruit salads are, therefore, often sold immersed in sugar syrups to increase shelf life. Pre-processing short-term hot-water treatments (sHWT) may further extend the shelf life of fresh-cuts by effectively reducing microbial contaminations before cutting. In this study, fresh-cut 'Braeburn' apples, a major component of fruit salads, were short-term (30 s) hot water-treated (55 °C or 65 °C), partially treated with a commercial anti-browning solution (ascorbic/citric acid) after cutting and, thereafter, stored immersed in sugar syrup. To, for the first time, comprehensively and comparatively evaluate the currently unexplored positive or negative effects of these treatments on fruit quality and shelf life, relevant parameters were analysed at defined intervals during storage at 4 °C for up to 13 d. Compared to acid pre-treated controls, sHWT significantly reduced the microbial loads of apple slices but did not affect their quality during the 5 d-standard shelf life period of fresh-cuts. Yeasts were most critical for shelf life of fresh-cut apples immersed in sugar syrup. The combination of sHWT and post-processing acid treatment did not further improve quality or extend shelf life. Although sHWT could not extend potential maximum shelf life beyond 10 d, results highlighted the potentials of this technique to replace pre-processing chemical treatments and, thus, to save valuable resources.

Keywords: minimal processing; sugar syrup immersion; microbial analyses; chemical prevention; ready-to eat fruit salads

1. Introduction

Processing of fresh-cut fruit salads involves sorting, cleaning, washing, peeling, deseeding/coring and cutting of fruit. Especially for apples, which are often a major component of fruit salads, cutting (Toivonen and DeEll, 2002) pronouncedly reduces the shelf life due to increased water losses (Beaudry, 2000; Watkins, 2000), surface browning and accelerated microbial spoilage (Artés et al., 2007; Capozzi et al., 2009). To optimize the shelf life of fresh-cut fruit, various postharvest treatments are used, alone and in combination (Rux et al., 2017), such as modified atmosphere and humidity packaging (Rux et al., 2015; Caleb et al., 2006), edible coatings (Rojas-Graü et al., 2007), thermal (Koukounaras et al., 2008) and plasma (Tappi et al., 2014) treatments, UV-C irradiation combined with modified atmosphere packaging (MAP) (López-Rubira et al., 2005) or the application of organic acid solutions and sugar syrups (Nishikawa et al., 2005). In practice, the latter is in particular used for bulk purchaser (Alzamora et al., 1993) to extend the common maximum shelf life for fresh-cut fruit salads of 5 d (Seide et al., 2017) to up to 10 d.

Despite reports on the effects of storage in sugar syrup on ethylene metabolism (Nishikawa et al., 2005) or chlorophyll degradation of fresh-cut fruit (Coupe et al., 2003), the knowledge on the impact of sugar solution on the physiological properties of immersed fresh-cuts is still limited. Investigating the effects of complete immersion of apple slices in sugar solutions of different concentrations on various physiological and quality aspects, Rux et al. (2017) indicated that concentrations between 13–20 % effectively prevented browning and showed best potentials for storage of fresh-cut fruit. The authors also reported that microbial spoilage was the main factor limiting shelf life of the apple slices. Consequently, careful removal of microorganisms, adherent to fruit skin (Pietrysiak and Ganjyal, 2018) is essential to avoid microbiological contamination and cross-contaminations (Kabelitz and Hassenberg, 2018).

Short-term (15 s up to few min) hot-water treatments (sHWT) at relatively high temperatures (40 – 80 °C) effectively reduce microbial contamination (Lurie, 1998) and insect infestations, and maintain storage quality of apples and other fruits (Trierweiler et al., 2003; Spadoni et al., 2015; Kabelitz and Hassenberg, 2018). In addition, sHWT is relatively inexpensive, easy to use and gentle, and needs no chemicals. Therefore, sHWT is particularly suitable for organic production (Fallik, 2004; Maxin et al., 2014).

For fruit salad production, sHWT needs to be optimized to prevent fruit injury and reduction of produce quality, and to guarantee the commercially required shelf life of 10 d. Although systematic studies on heat impacts on structure and function of fruit epidermal tissue are mainly restricted to long-term non-aqueous heat treatments at moderate temperatures (Roy et al., 1994; Lurie et al., 1996), Kabelitz and Hassenberg (2018) evaluated the suitability of sHWT for pre-processing of apples for fresh-cut fruit salad production. In addition, Kabelitz et al. (2019) investigated sHWT effects on the commodity surface tissue and heat transfer dynamics in "intact" apples. To the authors' best knowledge, the potential implications of sHWT on the physiological quality parameters of apple slices have not been studied yet under semi-practical conditions during and beyond the maximum shelf life periods (10 d and 13 d, respectively).

Thus, this study aims to evaluate the effects of short-term (30 s) HW-treatment (at 55 °C or 65 °C) on important physiological parameters of fresh-cut 'Braeburn' (commercially often used in fresh-cut fruit salads) apple slices, stored immersed in sugar syrup in pails for up to 13 d. sHW treatment conditions were chosen following Kabelitz et al. (2019), who showed that treatments at 55 °C optimally maintained fruit quality and efficiently reduced microbial loads, while those at 65 °C were optimal for pathogen inactivation. To simulate practical condition as close as possible, slices were partially treated with a commercial ascorbic/citric acid solution to prevent browning of surface after cutting. At defined intervals, samples were taken to analyse important quality parameters. Care was taken so that all treatments and processes reflect current practical use as close as possible.

For the first time, the currently unexplored positive or negative effects of sHWT and/or of post-processing acid treatment both combined with sugar syrup-storage on fruit quality and shelf life were comprehensively and comparatively evaluate on fresh-cut apple slices. The results of the presented study will allow assessing whether sHWT could potentially supplement or replace post-processing chemical treatments for improving quality maintenance and shelf life, the latter beyond the current maximum shelf life period of 10 d.

2. Materials and methods

2.1. Material

Fresh mature 'Braeburn' apples (*Malus domestica* Borkh.) were obtained from a commercial fresh-cut salad producer (mirontell fein & frisch AG, Großbeeren, Germany) and transported to the Department of Horticultural Engineering (Leibniz Institute for Agricultural Engineering and Bioeconomy, Potsdam, Germany). Undamaged apples of uniform size were selected and stored at 4 °C and 95 % relative humidity for up to 3 d until experiments. Average initial mass ($n = 20$) and dry matter content ($n = 6$) were 150.7 ± 5.1 g and 17.7 ± 1.1 g per 100 g fresh mass (FM), respectively.

2.2. Pre-Processing Short-Term Hot-Water Treatment

Following Kabelitz et al. (2019), a batch of 4 cooled apples (in total, 16 fruit per treatment) was either water-washed (controls) at room temperature (approx. 20 °C) or short-term hot-water treated at 55 °C or 65 °C for 30 s in a GFL 1086 water bath (Gesellschaft für Labortechnik mbH, Burgwedel, Germany) before cutting. During treatments, apples were kept submerged in water using a stainless steel plate. Before each additional treatment, the water temperature was readjusted and the water replaced after every second washing.

2.3. Fresh-Cut Preparation and Sampling

Closely following practical treatments, apples were further processed under hygienic conditions at 4 °C in a cooling room. Apples were halved equatorially, and each half-segment was cored and equally cut into 16 pieces using a Parti apple cutter (Gefu GmbH, Eslohe, Germany).

For the additional anti-browning treatments, the apple slices of some batches were immersed in ascorbic/citric acid solution (40 g ascorbic acid and 20 g citric acid in 1 L deionized water) for 5 min (**Table 5.1**), as it is currently applied in practice (mirontell fein & frisch AG, personal communication). Apple slices of each batch were carefully mixed to ensure homogeneous distribution and placed in a commercial 840 mL plastic pail (approx. 48 with ca. 230 g), with three pails for each sampling day as replicates, yielding a total of 12 pails per treatment.

Table 5.1. Treatments of apples slices, fresh (initial) or stored (at 4 °C; max. 13 d) in sugar syrup with or without chemical prevention.

Treatment	sHWT	cp	Immersed in sugar syrup	Sampling day
initial	no	no	no	0
control	no	yes	yes	5, 10, 13
55	55 °C	no		
55_cp	55 °C	yes		
65	65 °C	no		
65_cp	65 °C	yes		

sHWT = short-term hot-water treatment; cp = chemical prevention with organic acids.

All pails were filled with 450 mL sugar syrup (200 g L^{-1} invert sugar syrup; 72.7 %; Hanseatische Zuckerraffinerie GmbH and Co. KG, Hamburg, Germany) and 10 g L^{-1} OBSTSERVAL HC-2 (Konserval, Pharmacon Lebensmittelzusätze GmbH, Trittau, Germany) browning inhibitor (aqueous solution of ascorbic acid, sodium ascorbate and citric acid), tightly closed and stored at 4 °C for up to 13 d. On days 5, 10 and 13, three pails were opened and apple slices or aliquots of the syrup removed for further analyses. Days 5, 10 and 13 represent the common shelf life of fresh-cut fruit salads, the shelf life given by the producer for storage in sugar syrup and the shelf life potentially pro-longed by sHWT, respectively.

2.4. Colour Measurement

For each treatment sample, both peel surface and tissue colour parameters (CIE [Commission Internationale de l'Éclairage] L*, a*, and b*) of 10 apple slices per treatment were measured with a CM-2600d spectrophotometer (Konica Minolta Sensing Inc., Tokyo, Japan), calibrated against white and black tiles. The browning index, BI (Maskan, 2001), was calculated as:

$$\mathrm{BI} = \frac{(100(x - 0.31))}{0.17}$$

where, (1)

$$x = \frac{(\mathrm{a} + 1.75\mathrm{L})}{(5.645\mathrm{L} + \mathrm{a} - 3.012\mathrm{b})}$$

2.5. Tissue Strength

For each sample day, tissue strength of 10 apple slices per treatment was determined with a texture analyser (TA-XT Plus, Stable Micro Systems, Surrey, UK), fitted with a SMS-P/4 cylinder probe. Tissue strength was measured as maximum compression force (N) at 8 mm indentation (speed for test and post-test; 1 and 10 mm s^{-1}, respectively).

2.6. Total Soluble Solids, Total Titratable Acidity and Vitamin C Contents

Tissue juice was extracted from 10 apple slices per treatments using a garlic press and the juice centrifuged at 14500 rpm for 2 min using MiniSpin Plus (Eppendorf AG, Hamburg, Germany). The total soluble solids content (TSS; °Brix) was measured from aliquots with a DR301-95 electronic refractometer (Krüss Optronic, Hamburg, Germany), while total titratable acidity (TA) was obtained with an automated T50M Titrator (with Rondo 20 sample changer, Mettler Toledo, Gießen, Germany) by titration with 0.1 mol L^{-1} NaOH to pH 8.2. Vitamin C contents (mg per 100 g FM) were determined with a Reflectoquant® test kit (Merck, Darmstadt, Germany).

2.7. Microbial Analysis

For microbial analyses, 6 apple slices and approx. 10 mL sugar syrup was removed from each pail. Using the total plate count method, samples were evaluated in duplicate ($n = 2$) either on plate count agar (Carl Roth GmbH and Co. KG, Karlsruhe, Germany) for total aerobic mesophilic bacterial counts or on rose bengal chloramphenicol agar (Carl Roth GmbH and Co. KG) for yeast and mould counts, respectively. Apple samples (10 g) or syrup were transferred into sterile stomacher bags filled with 90 mL buffered peptone water and homogenized with a BagMixer® 400CC® lab blender (Interscience, Saint Nom, France) at speed 4 (10 strokes s^{-1}) for 2 min. Thereafter, the samples were serial diluted by adding 30 µL of each diluent into 270 µL of peptone salt solution in Rotilabo®-microtest plates (96er U-profile, Carl Roth GmbH and Co KG) and 100 µL from each dilution were pour-plated on respective growth media. Aerobic mesophilic bacterial counts were enumerated after 3 d at 30 °C, whereas yeast and mould were counted after 7 d at 25 °C, and results expressed as colony forming unit per gram (CFU g^{-1}).

2.8. Statistical Analysis

For each sample day and treatment, three pails were used as replicates. For colour and tissue strength evaluations, 10 individual apple slices from each pail were measured and their results averaged ($n = 3$). Similarly, for TSS, TA and vitamin C (in duplicate) analyses, 10 apple slices from each pail were juiced, measured and their results averaged ($n = 3$). Microbial loads were determined using 6 homogenized apple slices from each pail, which were analysed in duplicate and their results averaged ($n = 3$). All data were statistically analysed (ANOVA) using WinSTAT (R. Fitch Software, Staufen, Germany) and presented as means ± standard deviation (SD). Duncan's multiple range test ($p < 0.05$) was used to analyse the significance of differences between means.

3. Results

3.1. Colour Evaluation

Neither treatment conditions nor storage time clearly affected any colour parameter of both fruit tissue (**Table 5.2**) and peel (**Table 5.3**). Variations of L*, a*, b* or BI were small and did not reflect any clear trend. Nevertheless, lightness, L*, of stored tissue samples was significantly reduced irrespective of treatments, if compared to unprocessed fruit. Among sHWT samples, differences between peel colour parameters were not significant.

3.2. Tissue Strength

The mean initial tissue strength of freshly cut apples slices was 7.4 ± 1.4 N; it significantly and pronouncedly (>25 %) declined in all samples irrespective of the treatments during storage (**Fig. 5.1**). Softening, however, was largest in controls, where tissue strength declined to 3.7 N, i.e., by more than 50 % even during the initial 5 d of storage. Compared to controls, all sHWT-samples without additional application of acid solutions had significantly stronger tissue at day 5. Irrespective of treatments, tissue strength continued to decline by 10–32 % in all samples with further storage (10 and 13 d).

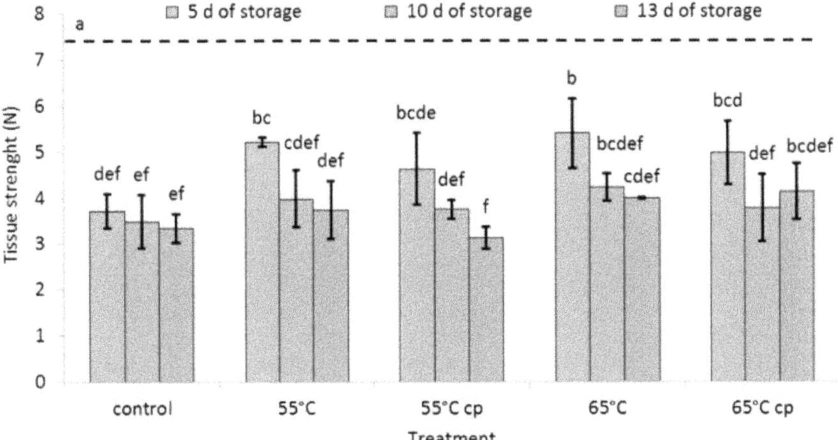

Figure 5.1. Tissue strength of short-term hot-water treated and untreated fresh-cut apple slices, stored at 4 °C in sugar syrup for up to 13 d compared to that of fresh fruit (dotted line). Given are means ± standard deviation ($n = 3$). Different letters indicate significant differences between means ($p < 0.05$).

Table 5.2. Tissue colour parameters (browning index (BI) and CIE L*, a*, b*) of short-term hot-water treated and untreated fresh-cut apple slices stored at 4 °C in sugar syrup for up to 13 d.

day	treatment	BI	L*	a*	b*
0	initial	35.3 ± 6.0 (a)	67.0 ± 1.7 (a)	2.8 ± 1.1 (a)	18.8 ± 2.5 (a)
5	control	32.5 ± 1.0 (a)	63.9 ± 0.4 (bc)	2.3 ± 0.2 (ab)	16.7 ± 0.2 (b)
	55	33.2 ± 1.8 (a)	65.1 ± 1.1 (b)	1.8 ± 0.0 (b)	17.7 ± 0.9 (ab)
	55_cp	35.0 ± 1.5 (a)	63.3 ± 0.8 (bc)	2.2 ± 0.3 (ab)	17.7 ± 0.5 (ab)
	65	34.3 ± 3.5 (a)	64.5 ± 0.7 (bc)	2.1 ± 0.3 (ab)	17.8 ± 1.4 (ab)
	65_cp	32.8 ± 0.9 (a)	64.5 ± 0.7 (bc)	2.1 ± 0.2 (ab)	17.2 ± 0.3 (ab)
10	control	31.2 ± 0.6 (a)	64.0 ± 0.1 (bc)	2.2 ± 0.3 (ab)	16.2 ± 0.2 (b)
	55	32.4 ± 1.6 (a)	64.6 ± 1.1 (bc)	2.2 ± 0.6 (ab)	17.0 ± 0.1 (ab)
	55_cp	31.7 ± 1.3 (a)	64.3 ± 1.2 (bc)	2.1 ± 0.1 (ab)	16.6 ± 0.3 (b)
	65	34.1 ± 3.6 (a)	64.7 ± 1.3 (bc)	2.5 ± 0.5 (ab)	17.6 ± 1.1 (ab)
	65_cp	33.7 ± 0.7 (a)	62.8 ± 1.0 (c)	2.4 ± 0.1 (ab)	16.9 ± 0.2 (ab)
13	control	33.2 ± 0.9 (a)	63.7 ± 0.5 (bc)	2.3 ± 0.2 (ab)	17.0 ± 0.6 (ab)
	55	33.6 ± 1.4 (a)	65.2 ± 1.4 (b)	2.3 ± 0.3 (ab)	17.6 ± 0.2 (ab)
	55_cp	32.8 ± 1.8 (a)	63.9 ± 1.4 (bc)	2.2 ± 0.1 (ab)	16.9 ± 1.1 (ab)
	65	34.7 ± 1.9 (a)	64.4 ± 0.6 (bc)	2.2 ± 0.2 (ab)	18.0 ± 1.1 (ab)
	65_cp	32.9 ± 1.6 (a)	64.3 ± 1.2 (bc)	2.2 ± 0.3 (ab)	17.1 ± 1.2 (ab)

Given are means ($n = 3$) ± standard deviation. Different letters indicate significant difference between means ($p < 0.05$). cp = chemical prevention with organic acids.

Table 5.3. Peel colour parameters (BI and CIE L*, a*, b*) of short-term hot-water treated and untreated fresh-cut apple slices stored at 4 °C in sugar syrup for up to 13 d.

day	treatment	BI	L*	a*	b*
0	initial	60.7 ± 6.3 (abc)	52.9 ± 2.5 (abc)	17.1 ± 3.3 (a)	16.6 ± 1.8 (a)
5	control	63.0 ± 2.1 (abcd)	53.0 ± 4.4 (abc)	13.9 ± 2.5 (ab)	19.1 ± 3.7 (ab)
	55	65.6 ± 1.7 (abcde)	55.6 ± 2.5 (bc)	12.4 ± 2.0 (bc)	21.9 ± 2.8 (bc)
	55_cp	61.8 ± 1.9 (abcd)	55.6 ± 2.1 (abc)	10.3 ± 1.4 (bcd)	21.5 ± 0.9 (bc)
	65	69.8 ± 4.7 (de)	57.3 ± 2.2 (c)	9.8 ± 3.3 (bcd)	25.3 ± 4.1 (cd)
	65_cp	65.5 ± 2.7 (abcde)	55.3 ± 0.5 (bc)	11.2 ± 0.7 (bcd)	22.3 ± 1.1 (bc)
10	control	59.3 ± 6.3 (ab)	55.0 ± 2.3 (ab)	12.0 ± 3.1 (bc)	19.5 ± 0.7 (ab)
	55	61.7 ± 3.7 (abcd)	56.5 ± 1.5 (abc)	10.2 ± 0.8 (bcd)	21.9 ± 1.4 (bc)
	55_cp	57.4 ± 4.6 (a)	58.6 ± 2.9 (a)	7.4 ± 2.5 (d)	22.7 ± 3.2 (bc)
	65	68.1 ± 7.3 (de)	57.7 ± 3.1 (c)	8.5 ± 2.4 (cd)	25.4 ± 0.6 (cd)
	65_cp	64.8 ± 5.4 (abcde)	57.0 ± 2.0 (abc)	8.2 ± 2.2 (cd)	24.2 ± 2.8 (cd)
13	control	65.3 ± 3.4 (abcde)	55.8 ± 2.6 (bc)	12.2 ± 2.0 (bc)	21.8 ± 1.5 (bc)
	55	66.6 ± 1.8 (bcde)	58.0 ± 1.0 (c)	8.8 ± 2.2 (cd)	25.0 ± 1.8 (cd)
	55_cp	63.8 ± 1.7 (abcde)	58.9 ± 1.2 (abc)	6.8 ± 2.3 (d)	25.4 ± 1.8 (cd)
	65	72.1 ± 3.1 (e)	57.9 ± 1.2 (c)	7.4 ± 0.8 (d)	27.4 ± 1.8 (d)
	65_cp	67.8 ± 2.1 (cde)	57.3 ± 3.0 (c)	9.3 ± 2.5 (cd)	24.9 ± 3.2 (cd)

Given are means ($n = 3$) ± standard deviation. Different letters indicate significant difference between means ($p < 0.05$). cp = chemical prevention with organic acids.

3.3. Total Soluble Solids, Total Titratable Acidity and Vitamin C

Irrespective of the treatments, the initial SSC of fresh apple slices (12.7 °Brix) did only marginally and insignificantly decrease during storage (**Fig. 5.2A**). Exclusively SSC of samples treated at 65 °C with additional acid treatment (65 cp) slightly declined to 11.9 °Brix at day 10.

In all samples, total titratable acidity (TA) increased during storage, starting from an initial TA of 0.39 mg 100 g^{-1} in fresh apples (**Fig. 5.2B**). In particular during initial and late storage, the rise in TA content was significant in most samples. During the entire storage, TA was not significantly different in the apple slices, irrespective of the treatments.

The initial vitamin C content of fresh apples (76 mg 100 g^{-1}) drastically increased during initial storage, in particular in samples additionally treated with acid solution (**Fig. 5.2C**). In the latter samples, the additional application of vitamin C at high concentrations (i.e., 40 g L^{-1}) to the freshly cut slices significantly increased their acid content compared to apples slices without this post-processing treatment (vitamin C content in sugar syrup < 10 g L^{-1}). In all samples, vitamin C contents tended to increase during further storage although at lower rates due to the reduced concentration gradients.

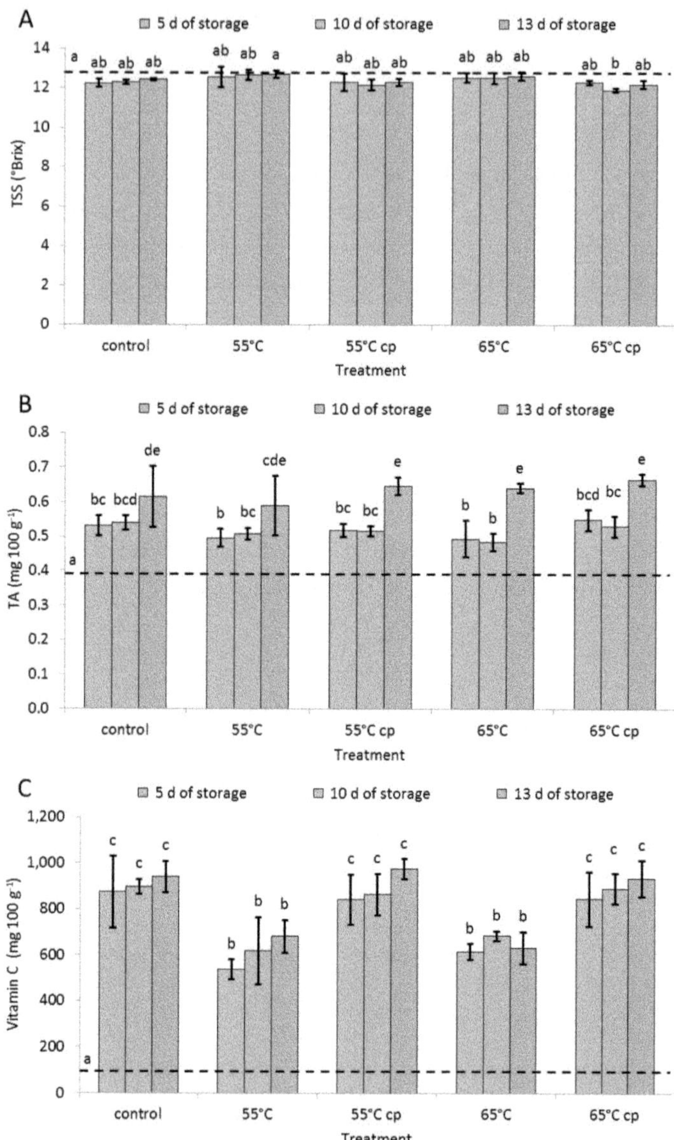

Figure 5.2. (**A**) Total soluble solids (TSS), (**B**) total titratable acidity (TA) and (**C**) vitamin C content of short-term hot-water treated and untreated fresh-cut apple slices stored at 4 °C in sugar syrup for up to 13 d compare to initial values measured in fresh fruit (dotted line). Given are means ± standard deviation ($n = 3$). Different letters indicate significant differences between means ($p < 0.05$).

3.4. Microbial Analysis

Compared to initial total aerobic mesophilic bacteria (TAMB) counts of fresh apples (6.4 Log CFU g^{-1}), TAMB counts were strongly reduced after treatments in all samples but partially increased again with prolonged storage (**Fig. 5.3A**). On day 5, there was no clear cut effect on TAMB amongst samples of the different treatments, although sHWT and acid-treated apples always showed slightly smaller counts. This effect was more pronounced after 10 d but not after 13 d of storage. The initial yeast counts (3.5 Log CFU g^{-1}) of fresh apples did not significantly change during early storage, but significantly increased with longer storage, irrespective of the treatment (**Fig. 5.3B**). In samples of all treatments, mould counts were significantly lower after treatments (**Fig. 5.3C**) than in fresh apples (initial counts: 2.2 Log CFU g^{-1}). Some sHWT samples (55 cp and 65) showed significantly lower mould counts than controls. From day 10 onwards, no differences between mould counts were observed irrespective of the treatments.

The TAMB in sugar syrup increased during storage irrespective of the treatments (**Fig. 5.4A**). Except for controls, this increase, however, was not significant. The TAMB load of the sugar syrup of apple slices treated at 65 °C and, additionally, with acid solution (65 cp) was significantly lower than that of the sugar syrup of controls. No significant growth of TAMB was observed in sugar syrup until day 10 of storage. However, syrup containing samples treated at 55 °C plus acid (55 cp) and at 65 °C without acid (65) had significant lower TAMB loads on day 10 if compared to that of controls. After prolonged storage, TAMB counts were significantly higher in control syrup and that of 55 °C sHWT samples, if compared to that on day 5.

No moulds could be detected in sugar syrups, irrespective of treatments and time of storage. In contrast, the low yeasts counts in clean sugar syrups (0.7 Log CFU g^{-1}) continued to significantly increase with storage for all syrups (**Fig. 5.4B**). Only on day 5, the sugar syrups that had contained sHWT samples showed significantly lower yeast counts than those containing controls. However, sugar syrups of 65 °C-treated apple slices yielded higher yeast counts than those of the other sHWT samples.

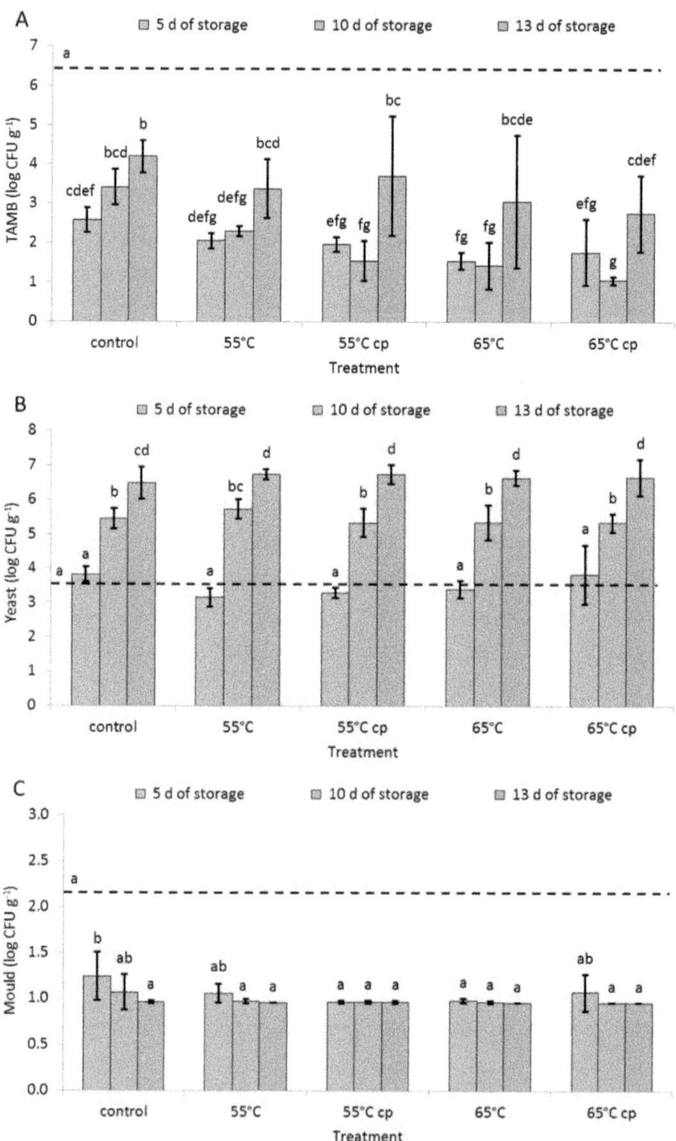

Figure 5.3. Microbial analysis of fresh, short-term hot-water treated and untreated fresh-cut apple slices stored at 4 °C in sugar syrup for up to 13 d. (**A**) Total aerobic mesophilic bacteria (TAMB), (**B**) yeast and (**C**) mould. The dotted lines represent the initial contamination. Given are means ± standard deviation ($n = 3$). Different letters indicate significant differences between means ($p < 0.05$).

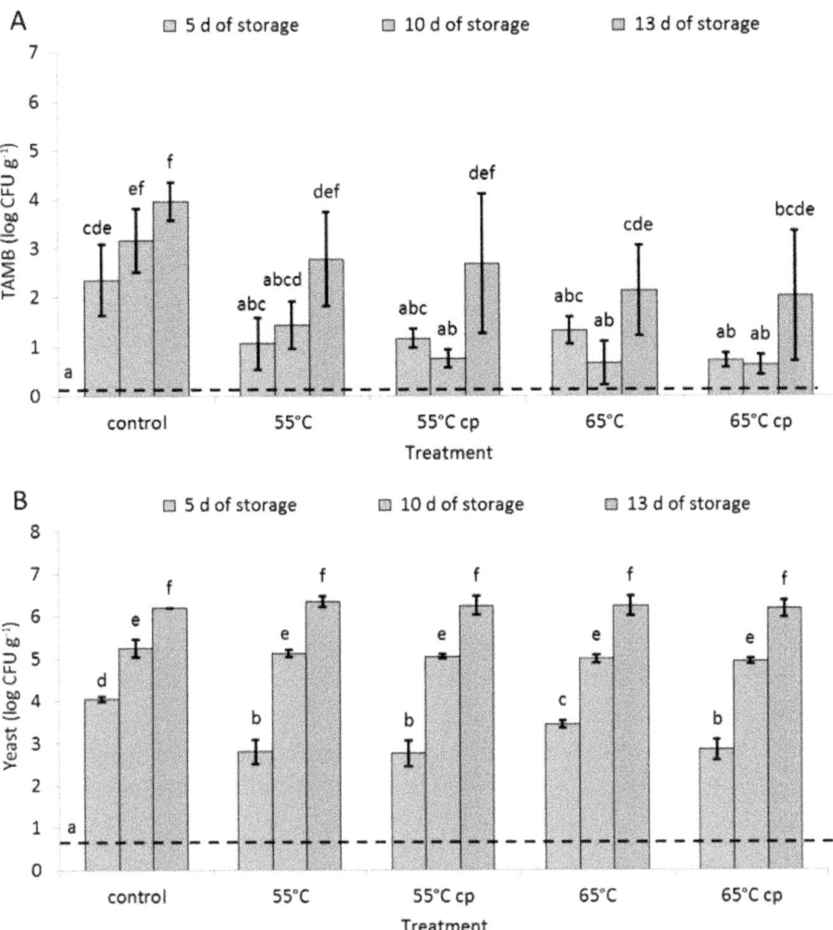

Figure 5.4. Microbial analysis of sugar syrup containing fresh, short-term hot-water treated or untreated fresh-cut apple slices stored at 4 °C for up to 13 d. (**A**) Total aerobic mesophilic bacteria (TAMB) and (**B**) yeast. The dotted lines represent the initial contamination. Given are means ± standard deviation ($n = 3$). Different letters indicate significant differences between means ($p < 0.05$).

4. Discussion

4.1. Tissue and Peel Colour

Surface colour is a very important quality parameter of fresh-cut fruit, which greatly impacts consumers' acceptance (Toivonen and Brummell, 2008). In particular, extensive surface browning of cut tissues may lead to the rejection of fresh-cut fruit salads.

In the present study, storage in the commercial sugar syrup, regularly containing ascorbic acid, sodium ascorbate and citric acid, effectively prevented tissue browning (indicated by the browning index, BI) in all samples. The additional acid treatments after cutting did not further enhance this and appeared to have no advantageous effect. Furthermore, sHWT did not significantly affect tissue colour of apple slices.

Tissue browning is closely related to the concentration of phenolic compounds and to the activity of the key enzyme polyphenol oxidase (Martinez and Whitaker, 1995), while tissue pH, temperature and especially the availability of oxygen also play important roles (Martinez and Whitaker, 1995). Consequently, reduced O_2 concentrations may prevent browning and, thus, the reduction of lightness L* as, e.g., shown for 'Golden Delicious' apple slices in 100 % N_2 (Soliva-Fortuny et al., 2002). Storage of apple slices in modified atmosphere packaging (MAP) at 0 – 1 kPa O_2 (Gunes and Watkins, 2001) or at 5 kPa O_2 + 5 kPa CO_2 (Rocculi et al., 2004) may even increase tissue lightness. Furthermore, storing fresh-cut apple slices immerged in commercial sugar syrups of different sugar concentrations completely prevented surface browning (Rux et al., 2017). This effect was attributed to both the reduced oxygen availability in the syrup (Rux et al., 2017) and to the addition of citric acid as active anti-browning agent (Cocci et al., 2006; Cortellino et al., 2013) to the sugar solution. The reduced pH in the syrup may further intensify the inhibition of enzymatic browning (Nicolas et al., 1994; Whitaker, 1994; Chiabrando and Giacalone, 2012).

In the present study, peel browning was not observed in any treatment. Apple slices treated at 65 °C, however, experienced a slightly reduction of the red peel colour due to the degradation of dissolved anthocyanins, while the yellow colour increased compared to controls. This might be mainly attributed to the fact that, in contrast to the cells of the fruit body tissue, the epidermis and few hypodermal cell layers were pronouncedly heated (Kabelitz et al., 2019). Treating apples at temperatures above 55 °C may cause necrotic damages and browning of the apple peel (Kabelitz and Hassenberg, 2018); whereas temperatures below 50 °C even prevented enzymatic browning due to inhibition of the polyphenol oxidases (Kim et al., 1993; Kurenda, et al., 2014).

4.2. Tissue Strength

Tissue strength is a very complex parameter (Herppich et al., 2004) resulting from multiple tissue and cell properties such as tissue structure, cell to cell adhesion, turgor, cell size, and biochemical and biophysical characteristics of cell walls (Toivonen and Brummell, 2008; Herppich et al., 2005). Tissue softening can be related to various physiological processes such as hydrolysis of protopectins to water-soluble pectins, diffusion of symplastic sugars into intercellular spaces, decrease in cellulose crystallinity, movement of ions out of the cell wall, or cell wall thinning (Toivonen and Brummell, 2008). In the present study, tissue strength significantly and more or less continuously declined during storage in sugar syrup. At the relatively mild temperature and the short incubation time used, sHWT did not accelerate softening but tended to diminish this effect. These findings are supported by earlier reports that heat-treatment may improve tissue strength of intact apples (Trierweiler et al., 2003; Fan et al., 2011). Inhibition of softening by sHWT was in particular obvious in samples not additionally treated with acid solutions. In this context, Ponting et al. (1972) and Rojas-Graü et al. (2007) also reported a decrease in the tissue strength in ascorbic acid treated apple slices. It can, thus, been assumed that these treatments may induce or facilitate acid related degradation of pectins (Zhu et al., 2007) and, concomitant, cell wall weakening, thus leading to softening. Rux et al. (2017), investigating the effects of different immersion media on quality properties of fresh-cut apple slices, indicated that the osmolality of syrups had the strongest influence on tissue softening. Immersion of samples in 20 % sugar solutions even slightly (9 %) enhanced tissue strength compared to fresh-cut samples. Nevertheless, in all the above mentioned studies, the choice of cultivars (Kim et al., 1993) and the stage of fruit maturity may certainly also play important roles (Gorny et al., 2000; Rojas-Graü et al., 2007).

4.3. Soluble Solid Content, Total Titratable Acidity and Vitamin C

Results of the present study clearly highlighted that neither sHWT temperature nor storage in sugar syrup affected the sugar contents (as indicated by TSS) of apple slices. This coincides with other investigations showing that storage of fresh-cut apple slices in air (6 d; Rux et al., 2017) or in low-oxygen atmosphere (at 4 °C for 12 d; Rocculi et al., 2004) did not significantly change the SSC of samples. On the other hand, Biegańska-Marecik and Czapski (2007) reported that pre-treating fresh-cut apple slices with different sucrose solutions before vacuum packing increased the SSC of samples depending on the respective sugar concentrations. In apple slices, stored in sugar solutions of various concentrations, the SSC gradually decreased (5 – 10 %) or increased (>20 %) according to the osmotic content of each solution (Rux et al., 2017). Thus, the concentration of sugar syrup (approx. 20 %) used in the present study was optimal and did not affect the sugar content, a very important component of fruit taste.

On the other hand, the sHWT at both temperatures had no (negative) effects on SSC, total titratable acids or vitamin C contents of fresh-cut apple slices. The increase in TA and in vitamin C observed during storage reflects the absorption of the certain amounts of ascorbic and citric acid, regularly added to the commercial sugar syrup. This is, in particular, obvious for the vitamin C contents of samples, additionally treated with the acid solution before storage. In this context, the ascorbic acid content of this acid solution is more than four times that of the commercial sugar syrup (see Materials and Methods). Similarly, Cocci et al. (2006) reported a significant increase in TA in apple slices after immersion in antioxidant solution (3 min, 1 % ascorbic acid + 1 % citric acid).

4.4. Microbial Load

In this study, sHWT and post-processing acid treatment, two approaches to reduce microbial contaminations, showed to work successfully for fresh-cut apple samples. Generally, the reduction mechanism of the post-cutting acid treatment is related to the antimicrobial effects of high acidity. This is the most important reason for the practical use of this approach. However, both techniques are not synergistic but resulted in similar reduction effects. Nevertheless, this also indicates the capability of sHWT to replace the chemical sanitation method. The processing industry is very interested in applying sHWT because it can lower the use of chemicals and may reduce production costs, although the energy expenditure must be considered.

In HWT, temperatures between 50 and 55 °C are generally assumed to effectively inactivate bacteria and yeasts (Trierweiler et al., 2003; Maxin et al., 2012a,b), the dominant microorganisms on intact apples (Chand-Goyal and Spotts, 1996; Kabelitz and Hassenberg, 2018). Spadoni et al. (2015) e.g., reported a 80 % reduction of total bacterial counts by HWT in the above mentioned temperature range. HWT of apples at 55 °C for 0.5 to 2 min successfully reduced inoculated bacteria, yeasts and moulds by 2–3 log.

In the present study, all treatments applied pronouncedly reduced the initially high TAMB load of fresh apples during early storage. Although sHWT sustainably lowered TAMB counts both on apple slices and in the syrup, a major effect may be ascribed to the acid treatment (Rux et al., 2017). On the other hand, immersion of apple slices in sugar syrup alone did not reduce microbial loads. Furthermore, sHWT at the higher temperature of 65 °C or the combination of both acid and sHWT only marginally improved hygienisation effects. Furthermore, sHWT of 65 °C appeared to shorten the shelf life of intact apples (Kabelitz and Hassenberg, 2018).

Yeast counts recorded for fresh apples and for apple slices after 5 d of storage in this study closely reflected results recently reported for intact apples (Kabelitz and Hassenberg, 2018). Although not obvious from the present results, acid- and/or sHWT presumably immediately reduced yeast counts, but continuous yeast growth led to identical counts on all treated samples after 5 d of storage. This duration of shelf life corresponds to the commonly demanded minimum shelf life for fresh-cut fruit salads (Seide et al., 2017). Only yeast counts exceeded the limits (5 log CFU g^{-1}) defined by the German Association for Hygiene and Microbiology (DGHM, 2011) for fresh-cut and packaged fruits after 10 d of storage. Therefore, in the present study, yeasts are most relevant for the maximum obtainable shelf life of fresh and fresh-cut apples immersed in sugar syrup. In case of yeasts, however, sHWT did only indirectly and temporally extended shelf life of fresh-cut fruit salads.

In this context, moulds did not cause any problem. Although sHWT somewhat intensified the positive effects, mould counts were reduced after all treatments and showed no further growth. This was probably due to the low pH combined with the high sugar contents in the syrup, which effectively suppressed any mould growth (Tapia de Daza et al., 1996).

The fact that overall microbial loads were not lower on slices of sHWT apples than on post-cutting acid treated controls may rule out the induction of protective physiological heat acclimation processes (Kabelitz et al., 2019). The above results, in particular the lack of any effect of the temperature rise from 55 °C to 65 °C, also substantiate the assumption that a direct heat inactivation of microorganisms is not the main mechanism of HWT sanitation (Kabelitz et al., 2019). On the other hand, in intact fruit, this temperature increase resulted in a sustained significant growth reduction of microbes during subsequent storage. These divergent responses of intact and sliced apples may indicate that the short-term and temporary melting of the epidermal wax layer, which produces a physical barrier to subsequent infections, may play an important role as also proposed by Kabelitz et al. (2019).

The microbial quality of storage syrups obviously also played a major role in the maintenance of quality and safety, because the sugar solutions presented a large proportion of the total mass (usually approx. 1/3) of stored fresh-cut fruit salads. In nearly aseptic pure syrups, all microbial counts were clearly below detection limits

(<1 Log). However, syrups may be contaminated by microorganisms adherent to the slices. Under sub-optimal storage conditions, particularly yeasts grow faster in the syrup than on apple slices, where they only colonize the surface. Thus, yeast counts in the entire pail may exceed legal limits, while that on apple slices may remain acceptable.

Due to the low pH in the solution, only TAMB and yeast counts increased during storage. Interestingly, sHWT of apples significantly inhibited the increase of yeast counts in the syrup during the initial 5 d of storage, although it had no effects on yeasts on fruit slices. In addition, yeast counts exceeded DGHM limits (5 log) in the sugar syrup on day 10 and thus proved their particular relevance in spoiling of fresh-cut fruit salads stored immersed in sugar syrup.

5. Conclusions

Pre-processing sHWT at both 55 °C and 65 °C did not adversely affect important quality parameters of fresh-cut apple slices and had no effects on TSS, total titratable acids or vitamin C contents. No peel or tissue browning could be observed. Although tissue strength continuously declined during storage in sugar syrup, sHWT tended to partially diminish softening. Thus, the use of sHWT eliminates the need for post-processing acid treatment. Bacteria and yeasts were the dominant microorganisms on apple slices and their counts increased during storage, both on apple slices and in the syrup. Short-term HWT significantly but only temporarily inhibited growth of TAMB and yeasts in the syrup during storage. However, sHWT could not extend the shelf life of apple slices. Increasing the treatment temperature from 55 °C to 65 °C only marginally improved TAMB reduction. In the control samples without sHWT, the observed reduction of microorganisms can be mainly ascribed to the post-cutting acid treatment. However, the combination of this acid and the sHWT did not or only minimally enhance the reduction of microbes. Moreover, the combination of treatments did not further improve quality maintenance. Yeasts are the most critical parameter for the shelf life of fresh and of fresh-cut apples immersed in sugar syrup because yeast counts exceeded accepted limits for fresh-cut and packaged fruit after 10 d of storage, both on the slices and in the syrup. The microbial quality of the syrup also plays a major role because these sugar solutions present a large proportion of the total mass of stored fresh-cut fruit salads. Summarizing, the presented results pointed out the ability of sHWT to replace actual commercial post-cutting acid preservation on fresh-cut apples.

6 Effects of Pre-Processing Hot-Water Treatment on Aroma Relevant VOCs of Fresh-Cut Apple Slices Stored in Sugar Syrup

Article IV

Guido Rux [1,2,*], Efecan Efe [1,2], Christian Ulrichs [2], Susanne Huyskens-Keil [2], Karin Hassenberg [1] and Werner B. Herppich [1]

[1] Department of Horticultural Engineering, Leibniz Institute for Agricultural Engineering and Bioeconomy (ATB), 14469 Potsdam, Germany

[2] Research Group Quality Dynamics/Postharvest Physiology, Division Urban Plant Ecophysiology, Humboldt-Universität zu Berlin, 14195 Berlin, Germany

* Correspondence: grux@atb-potsdam.de; Tel.: +49-331-5599-920

Received: 16 December 2019; Accepted 8 January 2020; Published: 10 January 2020

Abstract: In practice, fresh-cut fruit and fruit salads are currently stored submerged in sugar syrup (approx. 20 %) to prevent browning, to slow down physiological processes and to extend shelf life. To minimize browning and microbial spoilage, slices may also be dipped in a citric acid/ascorbic acid solution for 5 min before storage in sugar syrup. To prevent the use of chemicals in organic production, short-term (30 s) hot-water treatment (sHWT) may be an alternative for gentle sanitation. Currently, profound knowledge on the impact of both sugar solution and sHWT on aroma and physiological properties of immersed fresh-cuts is lacking. Aroma is a very important aspect of fruit quality and generated by a great variety of volatile organic compounds (VOCs). Thus, potential interactive effects of sHWT and sugar syrup storage on quality of fresh-cut apple slices were evaluated, focusing on processing-induced changes in VOCs profiles. Intact 'Braeburn' apples were sHW-treated at 55 °C and 65 °C for 30 s, sliced, partially treated with a commercial ascorbic/citric acid solution and slices stored in sugar syrup at 4 °C up to 13 d. Volatile emission, respiration and ethylene release were measured on storage days 5, 10 and 13. The impact of sHWT on VOCs was low while immersion and storage in sugar syrup had a much higher influence on aroma. sHWT did not negatively affect aroma quality of products and may replace acid dipping.

Keywords: minimal processing; sugar syrup immersion; volatile organic compounds; chemical prevention; ready-to-eat fruit salads

1. Introduction

Fresh-cut processing induces a catena of physiological responses (Watada and Qi, 1999; Beaulieu, 2010) finally resulting in the loss of quality and aroma and pronouncedly shortens the storage life of fresh-cut produce (Toivonen and DeEll, 2002). In current practice, fresh-cut fruit for fruit salads are often stored in sugar syrup (ca. 20 %), especially for use by bulk purchasers (Alzamora et al., 1993). This storage method may extend product shelf life by preventing enzymatic and oxidative browning and transpiration, slows down respiration, ethylene metabolism and other physiological processes (Nishikawa et al., 2005; Coupe et al., 2003; Rux et al., 2017). In this context, however, microbial spoilage is the main factor limiting shelf life (Rux et al., 2017). It is therefore very important to remove the microorganisms adherent to the fruit skin (Pietrysiak and Ganjyal, 2018) before processing.

In practice, e.g., apple slices are dipped in a mixture of citric and ascorbic acid solutions for sanitation purposes. To prevent the consumption of these chemicals especially in organic production, gentle physical sanitation methods are demanded.

As such, hot-water treatments (HWT) in the temperature range of 40 – 80 °C were shown to effectively reduce microbial contamination, and are relatively inexpensive and easy to use (Lurie, 1998). In addition, HWT maintains storage quality of fruits (Shao et al., 2007; Spadoni et al., 2015; Kabelitz and Hassenberg, 2018). Since they are chemical-free, particularly short-term (15 – 60 s) hot-water treatments (sHWT) are suitable for organic production (Fallik, 2004; Maxin et al., 2014). Besides earlier studies on the impacts of HWT on structure and function of fruit epidermal tissue (Roy et al., 1994; Lurie et al., 1996), only recently, the effects of sHWT on surface tissue, heat transfer dynamics and suitability for pre-processing of intact apples for fresh-cut salads were investigated in detail (Kabelitz and Hassenberg, 2018, 2019). Furthermore, the implications of sHWT on important quality parameters such as tissue browning, tissue strength and on microbial loads of apple slices immersed in sugar syrup have been studied under semi-practical conditions (Rux et al., 2019b). On the effects of sugar syrup storage on fresh-cut fruit quality attributes only very few studies are available (Biegańska-Marecik et al., 2007; Rux et al., 2017).

In contrast to analyses of sHWT impacts on visual and internal quality attributes of fresh-cut fruit (Kabelitz and Hassenberg, 2018; Kabelitz et al., 2019; Rux et al., 2019b), studies on potential variations in fruit aroma, a very important aspect of fruit quality sensation (Beaulieu, 2010; Barrett et al., 2010), are completely lacking. Aroma is generated by a great variety of permanent or secondary VOCs (Schwab et al., 2008), synthesized via numerous biosynthetic pathways, which, in turn, are regulated by a great variety of enzymes and substrates (Baldwin et al., 2000; Beaulieu, 2010; Espino-Díaz et al., 2016). Moreover, both respiration and ethylene biosynthesis are also involved in VOCs evolution (Schaffer et al., 2007). In addition, microbial growth may, directly or indirectly, negatively affect product aroma (Lamikanra et al., 2000; Rux et al., 2019a). Short-term-HWT and sugar syrup immersions may influence all of these processes. Knowledge on the impact of both sugar solution and sHWT on aroma development and physiological properties of immersed fresh-cuts is lacking.

Thus, the present study focused on the evaluation of the potential effects of sHWT (30 s, at 55 °C or 65 °C) on the aroma-related quality of 'Braeburn' apple slices stored in sugar syrup under strictly simulated practical condition. Some parts of the samples were also pre-treated with a commercial ascorbic/citric acid solution to test the potential synergistic effects of this treatment on aroma. During a 13 d-storage at 4 °C, respiration, ethylene emission and, for the first time, the processing-induced direct and indirect changes in VOC profiles were measured at defined intervals. This will enable comprehension of the respective quality-related physiological processes, to effectively select the optimal process conditions and to verify whether sHWT can safely replace the use of chemicals in processing of ecologically produced fresh-cut fruit salads.

2. Materials and methods

2.1. Material

Fresh mature 'Braeburn' apples (*Malus domestica* Borkh.) were obtained from a commercial fresh-cut salad producer. At the Department of Horticultural Engineering (Leibniz Institute for Agricultural Engineering and Bioeconomy, Potsdam, Germany), the apples were stored at 4 °C and 95 % relative humidity for up to 3 d until the start of the experiments. Undamaged apples of uniform size (mean fresh mass: 150.7 ± 5.1 g and mean dry matter content: 177 ± 11 g kg^{-1}) were selected.

2.2. Pre-Processing Short-Term Hot-Water Treatment

Before cutting, apples were divided in five batches of 16 fruit each. Apples from the control batch were water-washed at approx. 20 °C. The other samples were hot-water-treated in a GFL 1086 water bath (Gesellschaft für Labortechnik mbH, Burgwedel, Germany) by submerging four apples each for 30 s. Samples of two batches each were hot-water-treated at 55 °C or 65 °C, respectively, according to Kabelitz and Hassenberg (2018) and Kabelitz et al. (2019). These authors indicated that HWT at 55 °C for 30 s effectively reduced microbial loads without negatively affecting the external quality of samples, while 65 °C showed to be a negative control treatment in terms of quality maintenance.

2.3. Fresh-Cut Preparation and Sampling

After washing and hot-water treatments, all apples were cut under semi-practical hygienic conditions in a cooling room at 4 °C. For this, apples were at first halved equatorially and then each half-segment was cut into 16 pieces by a Parti apple cutter/corer (Gefu GmbH, Eslohe, Germany). The controls and slices of one of each 55 °C and 65 °C sHWT batches were additional chemically treated (= chemical prevention, cp) by immersing in ascorbic/citric acid solution (40 g ascorbic and 20 g citric acid solved in 1 L deionized water) for 5 min (**Fig. 6.1**). Slices of each batch were randomly filled in 12 commercial 840 mL plastic pails. Closely following practice, each pail containing approx. 48 apple slices (approx. 228 g) was filled up with 450 mL sugar syrup (200 g L^{-1} invert sugar syrup, 72.7 %; Hanseatische Zuckerraffinerie GmbH & Co. KG, Hamburg, Germany plus 10 g L^{-1} OBSTSERVAL HC-2 browning inhibitor (Konserval, Pharmacon Lebensmittelzusätze GmbH, Trittau, Germany). The browning inhibitor contained ascorbic acid, sodium ascorbate and citric acid. All pails were tightly closed and stored at 4 °C for up to 13 d. Three pails (replicates) of each batch were opened on days 5, 10 (the common maximum shelf life) and on day 13. For further analyses, 26 apple slices of each pail were removed.

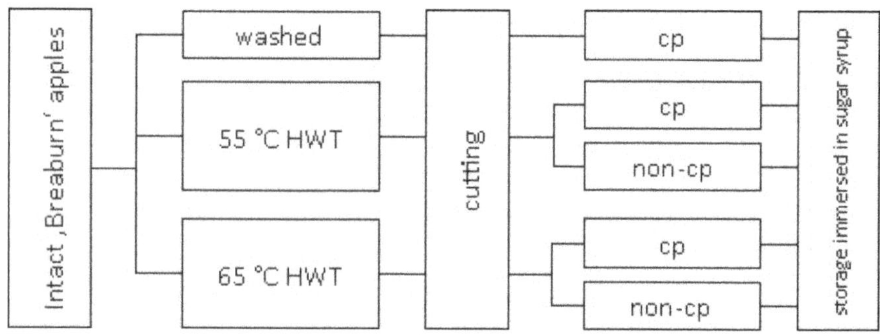

Figure 6.1. Scheme of the various pre- (short-term hot-water, sHWT, vs. washed) and post-cutting (chemical prevention, cp, vs. no chemical prevention, non-cp) treatments of fresh-cut apples slices stored in sugar syrup at 4 °C for up to 13 d.

2.4. Sampling of Volatile Organic Compounds (VOCs) and Ethylene

To determine VOCs emission, ethylene evolution and respiration rate (RR), 20 apple slices were filled in a glass jar (1 L), which was hermetically closed by a glass lid. VOCs emitted by samples accumulated at 4 °C for 1 h. Then VOCs were 10 min-extracted with 50/30 µm divinylbenzene/carboxen/polydimethylsiloxane (DVB/CAR/PDMS) SPME fiber (stableFlex/SS, Supelco, Bellefonte, PA, USA) via septum in glass lid. After VOCs extraction, 0.5 mL sample air was taken from headspace with a 0.5 mL A-2 Series syringe (VICI Precision Sampling, Baton Rouge, LA, USA) via the same septum to measure ethylene and CO_2 concentrations.

2.5. Analysis of Volatile Organic Compounds

VOCs were analysed with a GCMS-QP2010 SE gas chromatograph-mass spectrometer (GC–MS; Shimadzu Europa GmbH, Duisburg, Germany) on a DB-WAX (Agilent Technologies, Palo Alto, CA, USA) column (30 m × 0.25 mm × 0.25 µm). SPME fiber was introduced into the GC injector (set to 250 °C and fitted with a SPME liner (0.75 mm × 5.0 mm × 95 mm); Shimadzu GCs; Restek, Bellefonte, PA, USA) for 30 min for VOC desorption and conditioning before next extraction. Helium was used as carrier gas (flow rate: 0.8 mL min^{-1}). The following temperature program was used: 35 °C (hold 1 min), ramp to 110 °C at 5 °C min^{-1} and ramp to 230 °C at 10 °C min^{-1} (hold 3 min). The mass spectrometer (MS) detector worked in full scan mode (mass-to-charge ratio: 35–250 m/z) and operated in electron impact mode at 70 eV. Measured chromatogram peaks were integrated based on quantifier and qualifier ions of target compound (**Table 6.1**) using MSD Chem-Station E.02.01.1177 (Agilent Technologies, Palo Alto, CA, USA). Specific VOCs were identified (1) by NIST v2.0f library (NIST, Gaithersburg, MD, USA) and (2) by retention index (RI) calculated for each compound according to van Den Dool and Kratz (1963). RIs were determined based on an alkane standard (C7–C30; 1000 µg mL^{-1}; Supelco, Bellefonte, PA, USA). For this, 5 µL of alkane standard was injected in a 3.9 L glass jar, extracted with SPME after equilibrium time of 1 d and analysed as described above. The RIs were compared with those given by van Den Dool and Kratz (1963) by using online NIST Chemistry WebBook (2018). The total integrated peak area of each VOC was used to calculate a semi-quantitative VOC concentration expressed in nL L^{-1} based on the external standard 2-methylbutyl 3-methylbutanoate (CAS-number: 2245-77-4). A standard curve of 2-methylbutyl-3-methylbutanoate was created in the concentration range from 5 to 1000 nL L^{-1}.

Table 6.1. All identified volatile organic compounds (VOCs) emitted by intact, freshly cut and variously treated (sHWT/chemical prevention) fresh-cut apple slices as characterized by their CAS registry numbers, calculated retention indices (RI), previously reported retention indexes ($RI_{literature}$), and quantifier and qualifier ions used for peak integration.

Chemical Group	VOCs	CAS-Number	RI	$RI_{literature}$	Quantifier (Qualifier) Ions (m z^{-1})
esters (21)	Ethyl acetate	141-78-6	903	863–908	43 (45–61)
	Ethyl propionate	105-37-3	963	939–976	57 (74–75)
	Ethyl 2-methylpropanoate	97-62-1	968	957–969	43 (41–71)
	Propyl acetate	109-60-4	977	952–996	43 (61–73)
	Methyl butyrate	623-42-7	988	969–993	74 (43–71)
	Methyl 2-methylbutyrate	868-57-5	1011	1000–1010	57 (41–88)
	Isobutyl acetate	110-19-0	1014	1000–1031	43 (56–73)
	Ethyl butyrate	105-54-4	1035	1000–1073	71 (43–88)
	Ethyl 2-methylbutanoate	7452-79-1	1050	1022–1073	57 (41–102)
	Butyl acetate	123-86-4	1069	1049–1105	43 (56–73)
	2-methylbutyl acetate	624-41-9	1116	1111–1125	43 (55–70)
	Ethyl valerate	539-82-2	1129	1131–1139	88 (57–85)
	Ethyl 2-butenoate	10544-63-5	1154	1158–1158	69 (41–99)
	Pentyl acetate	628-63-7	1165	1175–1181	43 (55–70)
	Methyl hexanoate	106-70-7	1178	1176–1189	74 (43–99)
	Ethyl hexanoate	123-66-0	1222	1196–1245	71 (43–89)
	Hexyl acetate	142-92-7	1258	1251–1311	43 (56–61)
	2-Hexen-1-yl acetate	10094-40-3	1317	-	
	Hexyl butyrate	2639-63-6	1397	1393–1410	71 (43–89)
	Hexyl 2-methylbutanoate	10032-15-2	1414	1415–1416	
	Hexyl hexanoate	6378-65-0	1610	1596–1599	43 (56–117)
ketones (2)	2-Butanone	78-93-3	908	875–926	43 (57–72)
	1-Penten-3-one	1629-58-9	1020	1019–1056	
alcohols (6)	Ethanol	64-17-5	943	900–955	45 (43–46)
	2-Methyl-1-propanol	78-83-1	1096	1092–1114	43 (41–42)
	1-Butanol	71-36-3	1143	1116–1166	
	2-Methyl-1-butanol	1565-80-6	1200	-	57 (41–56)
	1-Hexanol	111-27-3	1341	1339–1396	
	2-(2-Ethoxyethoxy)-ethanol	111-90-0	1628	1615–1619	45 (59–72)
aldehydes	Hexanal	66-25-1	1077	1048–1120	44 (41–56)
terpenes (2)	D-Limonene	5989-27-5	1186	1176–1238	68 (67–93)
	α-Farnesene	502-61-4	1747	1720–1764	93 (41–69)
benzenic derivatives	Estragole	140-67-0	1724	1624–1661	148 (117–147)

2.6. Quantification of Ethylene and CO_2 Evolution

A small volume of 0.5 mL air was taken from canning glass jar and analysed with a GC17A gas chromatograph (Shimadzu) to measure ethylene and CO_2 concentrations. The GC was equipped with both FID and TCD, 80/100 Porapak N (Supelco, Bellefonte, PA, USA) column (1.8 m × 1/8 in × 2.1 mm) and a 5605PC mole sieve (Alltech GmbH, Unterhaching, Germany). As carrier gas, helium was used (constant flow rate: 1.4 mL min^{-1}), oven temperature was kept constant at 60 °C. Measured peaks were integrated using Class-VP chromatography data system software v4.2 (Shimadzu Europa GmbH, Duisburg, Germany). Rates of ethylene release were expressed in $\mu g\ kg^{-1}\ h^{-1}$, those of CO_2 production in $mg\ kg^{-1}\ h^{-1}$.

2.7. Statistical Analysis

For each treatment, three pails were used as replicates for each sample day. VOCs and ethylene emissions were each determined on 20 apple slices from each pail ($n = 3$), while respiration was analysed on a total of 60 apple slices, i.e., 20 from each pail. Statistical analyses (ANOVA) were carried out with WinSTAT (R. Fitch Software, Staufen, Germany) and results were presented as mean ± standard deviation (SD). The significance of differences between the calculated means was analysed by Duncan's multiple range test ($p < 0.05$). The relationships between VOC evolution and sHWT were additionally analysed by principal components analysis (PCA) using Latentix Ver. 2.00 (Latent5, Copenhagen, Denmark).

3. Results

3.1. Volatile Organic Compounds

In total, 33 different VOCs were identified in fresh-cut apple slices, with esters (21) forming the major group of compounds (**Table 6.1**). The emission of individual VOCs was significantly affected by cutting, sHWT and/or storage time (**Table 6.2**). In fresh, intact 'Braeburn' apples, 2-methylbutyl acetate (29.0 %) and α-Farnesene (15.9 %) were the most abundant VOCs. Cutting immediately and significantly increased the emissions of 17 VOCs, while that of only three VOCs significantly decreased. The cumulative VOCs concentration increased sevenfold, particularly concentrations of acetate esters (es2–es8) increased most.

Table 6.2. Semi-quantitative concentrations (nL L^{-1}) of VOCs, emitted by intact, freshly cut (cut) and variously treated (C: control with chemical prevention; 55: 55 °C sHWT without chemical prevention; 55_cp: 55 °C sHWT with chemical prevention; 65: 65 °C sHWT without chemical prevention; 65_cp: 65 °C sHWT with chemical prevention) fresh-cut apple slices on days 5, 10 and 13 of storage in sugar syrup at 4 °C. VOCs concentrations given resulted from the emissions of 20 apple slices hermetically enclosed in a 1 L-glass jar at 4 °C for 1 h. Given are means ($n = 3$). Different letters indicate significant differences between means ($p < 0.05$). The different colours help to indicate increasing (green) or decreasing (red) emissions of VOCs compared to intact apples (yellow).

VOC	Time	Initial	Time	C	55	55_cp	65	65_cp
	intact	87 a	5 d	445 b–d	515 c–e	388 b	415 b–c	425 b–c
Cumulative VOC concentration	cut	618 e	10 d	364 b	493 c–e	320 b	379 b	354 b
			13 d	385 b	502 d–e	305 b	317 b	369 b
Ethyl acetate	intact	0.86 a	5 d	109 c–d	152 e	80.9 b–c	93.3 b–d	83.5 b–c
es1	cut	1.98 a	10 d	104 b–d	180 f	75.8 b	121 d	83.2 b–c
			13 d	121 d	196 f	82 b–c	106 b–d	105 b–d
Propyl acetate	intact	nd	5 d	3.79 g	2.53 d–e	2.45 c–e	1.78 b–c	2.59 d–e
es2	cut	2.91 e–f	10 d	3.35 f–g	2.27 b–e	2.31 b–e	1.71 b	2.39 b–e
			13 d	3.74 g	2.21 b–e	2.11 b–d	0.39 a	nd
Butyl acetate *	intact	2.53 a	5 d	27.6 b	25.3 b	24.4 b	26.3 b	24.7 b
es3	cut	56.0 c	10 d	20.0 b	18.7 b	18.4 b	17.7 b	17.8 b
			13 d	19.0 b	17.3 b	16.9 b	14.9 b	17.2 b
Isobutyl acetate	intact	1.52 a	5 d	21.9 b	18.1 b	17.9 b	22.2 b	17.9 b
es4	cut	64.1 c	10 d	16.7 a–b	15.1 a–b	13.7 a–b	16.1 a–b	13.7 a–b
			13 d	15.9 a–b	15.8 a–b	12.3 a–b	13.9 a–b	14.2 a–b
2-methylbutyl acetate *	intact	25.2 a	5 d	139 d	127 c–d	110 b–d	114 b–d	128 c–d
es5	cut	323 e	10 d	94.1 b–d	94.7 b–d	79.5 a–c	74.8 a–c	91.1 b–d
			13 d	87.3 b–d	89.9 b–d	66.5 a–b	62 a–b	87.8 b–d
Pentyl acetate *	intact	0.78 a	5 d	4.36 b	3.34 a–b	3.36 a–b	4.04 b	3.16 a–b
es6	cut	13.1 c	10 d	2.64 a–b	2.44 a–b	2.02 a–b	2.37 a–b	1.91 a–b
			13 d	2.38 a–b	1.83 a–b	1.61 a–b	1.80 a–b	1.43 a–b
Hexyl acetate *	intact	3.15 a	5 d	5.98 a	4.64 a	4.45 a	5.19 a	3.73 a
es7	cut	59.8 b	10 d	3.48 a	3.56 a	2.31 a	2.85 a	2.14 a
			13 d	3.15 a	2.56 a	1.73 a	2.13 a	1.88 a
2-Hexen-1-yl acetate	intact	nd	5 d	nd	nd	nd	nd	nd
es8	cut	22.9 b	10 d	nd	nd	nd	nd	nd
			13 d	nd	nd	nd	nd	nd
Ethyl propionate	intact	1.46 b–c	5 d	3.17 e	3.16 e	2.03 c–d	1.22 b	1.65 b–d
es9	cut	nd	10 d	3.27 e	4.23 f	2.34 d	1.63 b–d	1.79 b–d
			13 d	5.01 g	4.49 f–g	2.33 d	1.34 b–c	1.82 b–d
Ethyl 2-methylpropanoate	intact	nd	5 d	1.06 b–c	2.88 e	1.32 b–c	1.64 c–d	1.24 b–c
es10	cut	nd	10 d	0.97 b–c	3.94 f	1.18 b–c	2.20 d–e	1.31 b–c
			13 d	1.35 b–c	4.76 g	0.63 a–b	0.82 a–c	0.89 a–c
Methyl butyrate	intact	0.63 a	5 d	1.37 a–b	1.06 a–b	1.39 a–b	0.77 a–b	1.79 b
es11	cut	4.20 c	10 d	1.13 a–b	0.93 a–b	1.36 a–b	0.63 a	1.48 a–b
			13 d	1.06 a–b	0.93 a–b	1.14 a–b	0.57 a	1.5 a–b
Methyl 2-methylbutyrate *	intact	nd	5 d	2.27 c–e	1.60 b–c	2.78 d–g	1.09 b	2.90 e–g
es12	cut	nd	10 d	1.97 b–d	1.62 b–c	3.48 g	1.09 b	3.27 f–g
			13 d	2.44 c–f	1.80 b–c	2.92 e–g	1.06 b	2.94 e–g
Ethyl butyrate *	intact	nd	5 d	12.1 b–c	16.8 c–d	12.4 b–c	13.4 b–c	12.5 b–c
es13	cut	1.64 a	10 d	10.9 b	18.4 d	11.9 b	13.3 b–c	11.6 b
			13 d	12.9 b–c	18.2 d	11.0 b	10.2 b	12.3 b–c

Table 6.2. *Cont.*

VOC	Time	Initial	Time	C	55	55_cp	65	65_cp
Ethyl 2-butenoate	intact	**nd**	5 d	**0.54 b–d**	**0.62 b–e**	**0.51 b–c**	**0.91 e–f**	**0.93 e–f**
es14	cut	**nd**	10 d	**0.70 b–f**	**0.82 c–f**	**0.55 b–d**	**0.98 f**	**0.78 b–f**
			13 d	**0.84 d–f**	**0.98 f**	**0.49 b**	**0.79 b–f**	**nd**
Ethyl 2-methylbutanoate *	intact	**nd**	5 d	**25.1 b–c**	**37.9 d–e**	**30.4 b–d**	**23.3 b**	**31.9 b–d**
es15	cut	**0.98 a**	10 d	**25.1 b–c**	**43.3 e–f**	**29.6 b–d**	**33.6 c–d**	**32.8 b–d**
			13 d	**33.7 c–d**	**51.0 f**	**27.4 b–c**	**24.8 b–c**	**33.4 c–d**
Hexyl butyrate *	intact	**5.24 a**	5 d	**1.55 b–d**	**1.92 b–d**	**2.45 b–c**	**2.66 b**	**2.51 b–c**
es16	cut	**5.39 a**	10 d	**1.00 d–e**	**1.16 c–e**	**1.44 b–e**	**1.38 b–e**	**1.65 b–d**
			13 d	**0.63 d–e**	**nd**	**nd**	**nd**	**1.28 c–e**
Hexyl 2-methylbutanoate *	intact	**5.56 a**	5 d	**11 b–e**	**12.9 e**	**12.6 d–e**	**11.3 c–e**	**11.2 c–e**
es17	cut	**8.20 b**	10 d	**11.2 c–e**	**11.1 b–e**	**10.2 b–e**	**9.76 b–d**	**12.3 d–e**
			13 d	**11.4 c–e**	**9.96 b–d**	**9.80 b–d**	**8.67 b–c**	**12.1 d–e**
Ethyl valerate	intact	**nd**	5 d	**0.37 d–e**	**0.47 f–g**	**nd**	**0.39 e–f**	**nd**
es18	cut	**nd**	10 d	**0.34 c–e**	**0.54 g**	**0.26 b–c**	**0.28 b–d**	**nd**
			13 d	**0.38 e**	**0.48 g**	**0.21 b**	**nd**	**0.3 b–e**
Methyl hexanoate	intact	**nd**	5 d	**0.58 b–d**	**1.80 e–f**	**0.54 a–d**	**1.44 e**	**1.01 d**
es19	cut	**2.09 f**	10 d	**0.41 a–b**	**1.89 e–f**	**0.39 a–b**	**0.98 c–d**	**0.44 a–b**
			13 d	**0.49 a–c**	**2.17 f**	**nd**	**nd**	**nd**
Ethyl hexanoate	intact	**nd**	5 d	**13.3 i–j**	**15.7 j**	**9.95 g–h**	**8.04 f–g**	**7.67 e–g**
es20	cut	**1.56 a–b**	10 d	**7.15 d–g**	**11.6 h–i**	**4.88 c–e**	**4.74 b–e**	**4.28 b–d**
			13 d	**6.34 d–f**	**7.72 e–g**	**2.86 a–c**	**2.74 a–c**	**3.22 a–c**
Hexyl hexanoate *	intact	**2.11 a–b**	5 d	**0.98 d–g**	**1.28 c–g**	**1.57 b–e**	**2.48 a**	**1.75 a–d**
es21	cut	**1.92 a–c**	10 d	**0.71 f–g**	**1.07 d–g**	**0.82 e–g**	**1.27 c–g**	**1.43 b–f**
			13 d	**0.56 g**	**0.71 f–g**	**0.73 f–g**	**0.92 e–g**	**1.14 c–g**
Ethanol	intact	**9.72 b**	5 d	**13.1 c**	**27.7 f**	**14.9 c**	**22.0 d–e**	**20.5 d**
al1	cut	**2.21 a**	10 d	**13.1 c**	**27.8 f**	**14.0 c**	**20.9 d–e**	**15.6 c**
			13 d	**15.0 c**	**32.4 g**	**19.7 d**	**23.8 e**	**20.1 d**
2-(2-Ethoxyethoxy)ethanol	intact	**0.65 a**	5 d	**0.35 b–d**	**0.45 b**	**0.43 b**	**0.37 b–d**	**0.38 b–c**
al2	cut	**0.62 a**	10 d	**0.30 c–e**	**0.25 d–e**	**0.22 e**	**0.29 c–e**	**0.22 e**
			13 d	**0.26 c–e**	**0.25 d–e**	**0.28 c–e**	**0.27 c–e**	**0.29 c–e**
2-Methyl-1-propanol	intact	**0.86 a**	5 d	**1.02 a–b**	**0.96 a**	**0.94 a**	**1.21 a–c**	**1.13 a–b**
al3	cut	**2.99 e**	10 d	**0.92 a**	**1.01 a–b**	**0.88 a**	**1.07 a–b**	**0.93 a**
			13 d	**1.07 a–b**	**1.68 d**	**1.50 c–d**	**1.36 b–d**	**1.09 a–b**
1-Butanol *	intact	**1.51 a**	5 d	**1.56 a**	**2.01 b–c**	**1.72 a–b**	**2.13 c**	**1.99 b–c**
al4	cut	**2.82 d**	10 d	**1.59 a**	**1.82 a–c**	**1.72 a–b**	**1.79 a–c**	**1.66 a–b**
			13 d	**1.46 a**	**1.67 a–b**	**1.67 a–b**	**1.67 a–b**	**1.64 a–b**
2-Methyl-1-butanol	intact	**3.24 a**	5 d	**4.25 a–b**	**5.32 b–d**	**4.44 a–c**	**5.14 b–d**	**5.89 c–d**
al5	cut	**9.02 e**	10 d	**4.63 a–c**	**5.58 b–d**	**4.67 b–c**	**4.60 a–c**	**5.21 b–d**
			13 d	**4.51 a–c**	**6.3 d**	**4.77 b–c**	**4.67 b–c**	**5.55 b–d**
1-Hexanol *	intact	**1.44 a**	5 d	**2.90 b**	**3.99 b–c**	**3.07 b**	**4.54 c**	**3.60 b–c**
al6	cut	**3.94 b–c**	10 d	**3.06 b**	**4.21 b–c**	**3.08 b**	**3.39 b–c**	**2.84 b**
			13 d	**2.88 b**	**3.93 b–c**	**3.27 b–c**	**3.52 b–c**	**3.11 b**
2-Butanone	intact	**2.67 a–e**	5 d	**0.86 a–b**	**4.71 e–f**	**3.62 c–e**	**8.89 g**	**6.75 f–g**
k1	cut	**0.33 a**	10 d	**0.73 a–b**	**7.32 g**	**1.96 a–d**	**4.68 e–f**	**4.22 d–e**
			13 d	**1.32 a–c**	**7.12 g**	**2.44 a–e**	**3.30 b–e**	**3.76 c–e**
1-Penten-3-one	intact	**nd**	5 d	**1.35 b**	**2.29 b–d**	**1.77 b–c**	**3.35 e–f**	**1.98 b–c**
k2	cut	**nd**	10 d	**3.71 e–f**	**3.98 f**	**3.21 d–f**	**5.16 g**	**3.72 e–f**
			13 d	**3.52 e–f**	**2.78 c–e**	**4.18 f**	**4.06 f**	**3.76 e–f**
Hexanal *	intact	**0.50 a**	5 d	**3.23 b–d**	**6.47 f–g**	**4.66 d–f**	**7.14 g**	**5.25 d–g**
ad1	cut	**1.92 a–b**	10 d	**4.49 d–f**	**5.08 d–f**	**3.72 b–d**	**6.26 e–g**	**4.43 d–e**
			13 d	**4.19 c–d**	**2.30 a–c**	**3.46 b–d**	**3.77 b–d**	**3.81 b–d**

Table 6.2. *Cont.*

VOC	Time	Initial	Time	C	55	55_cp	65	65_cp
Estragole	intact	0.30 b–e	5 d	0.61 g	0.42 c–f	0.45 d–g	0.35 b–f	0.47 e–g
bd1	cut	0.41 b–f	10 d	0.52 f–g	nd	nd	0.29 b–d	0.37 b–f
			13 d	0.51 f–g	nd	0.26 b–c	0.24 b	0.36 b–f
D-Limonene	intact	3.27 a	5 d	6.26 a–d	10.4 e	9.77 e	7.54 b–e	8.45 d–e
tp1	cut	4.93 a–c	10 d	4.74 a–b	4.54 a	9.54 e	7.65 c–e	8.67 d–e
			13 d	4.70 a–b	4.40 a	9.06 d–e	5.23 a–c	9.28 e
α-Farnesene	intact	13.8 a–c	5 d	24.0 d–e	18.8 a–d	21.3 c–e	16.8 a–d	28.0 e
tp2	cut	19.1 a–d	10 d	17.4 a–d	14.4 a–c	14.9 a–c	15.0 a–c	20.5 b–e
			13 d	16.5 a–d	10.6 a	12.1 a–b	12.5 a–b	17.7 a–d

* marked VOCs described as generally important "character impact" compounds in apples (Dixon and Hewett 2000). nd = not detected/below detection limit; dark green = ≥500 % increase; light green = ≥33 % increase; yellow = ≤33 % increase/decrease ; light red = ≥80 % decrease; dark red = ≥500 % decrease.

During further storage, the emissions of the majority of VOCs were significantly reduced compared to those observed immediately after cutting. Cumulative VOCs concentrations decreased by 37.7 % for control and by 18.8 – 50.7 % for sHWT samples, but were nevertheless 3.5 – 5.8 times higher than for intact apples (**Table 6.2**, 1st position). While cutting immediately and strongly increased the emission of acetate esters, the concentrations of these compounds decreased or, at least, remained constant during storage. Only the concentration of ethyl-acetate continued to increase after 3 d of storage as did the concentrations of all other ethyl esters, methyl 2-methylbutyrate, ethanol, hexanal and both ketones.

Responses of sHWT samples depended on both the treatment temperatures and the post-processing acid treatment (**Table 6.3**). Head space concentration of ethanol was generally higher and that of estragole and propyl-acetate lower in sHW-treated samples than in controls; propyl acetate concentration was more reduced at 65 °C than at 55 °C. Furthermore, sHWT at 65 °C significantly reduced emissions of ethyl-propionate, ethyl hexanoate and methyl hexanoate. These VOCs were also reduced at sHWT of 55 °C combined with acid treatment, i.e. chemical prevention. The combination of sHWT and acid application also resulted in increased D-Limonene and decreased ethyl 2-butenoate emissions, while the enhancement of the ethanol emission was attenuated. At a sHWT of 55 °C, the combination with organic acid resulted in lower ethyl acetate emission. In contrast, only the sole sHWT at 55 °C (i.e., without cp) let the emissions of nearly all ethyl esters increase especially that of ethyl-acetate, methyl hexanoate and 2-butanone, compared to controls.

Table 6.3. Treatment-specific qualitative changes in the emissions of relevant VOCs and corresponding aroma threshold values of fresh-cut apple slices stored in sugar syrup at 4 °C for up to 13 d.

Treatment	Reaction	VOCs	Aroma Threshold Values (nL L^{-1})
all sHWT	↑	Ethanol	8–900 [a]/>1 × 10^{5} [b]
	↓	Propyl acetate	2000–11,000 [a,b]
		Estragole	n/a
sHWT without cp	↑	Ethanol	8–900 [a]/>1 × 10^{5} [b]
sHWT combined with cp	↑	D-limonene	4–229 [a]
	↓	Ethyl 2-butenoate	n/a
only 55 °C sHWT	↑	2-methyl-1-propanol	360–3300 [a]
only 55 °C sHWT without cp	↑	Ethyl acetate	5–13,500 [b]
		Ethyl 2-methylpropanoate	0.01–1 [a]
		Ethyl butyrate *	0.1–18 [a]
		Ethyl 2-methylbutanoate *	0.006–0.1 [a,b]
		Ethyl valerate	1.5–5 [a,b]
		Methyl hexanoate	10–87 [a]
		2-butanone	n/a
	↓	Estragole	n/a
		1-penten-3-one	400 [a]
only 55 °C sHWT combined with cp	↓	Ethyl acetate	5–13,500 [a,b]
only 65 °C sHWT	↓	Propyl acetate	2000–11,000 [a,b]
only sHWT at 65 °C or combined with cp	↓	Ethyl propionate	9–45 [a]
		Ethyl hexanoate	0.3–5 [a]
		Methyl hexanoate	10–87 [a]

sHWT: short-term hot water treatment; cp: chemical prevention by applying organic acid dipping of fresh-cut slices; 55 °C/65 °C: sHWT at the specified temperature; green ↑: increase; red ↓: decrease. Given are aroma threshold values according to [a] Burdock (2016) and [b] Dixon and Hewett (2000). * marked VOCs described as generally important "character impact" compounds in apples (Dixon and Hewett, 2000).

The PCA provides two principal components (PC), which together represent 98.2 % of the total variance in VOCs profiles. The effects of cutting, sHWT temperature and duration of storage can be visualized by a scatter point plot (**Fig. 6.2A**). Decreasing values of PC1 (representing 65.3 % of total variance in VOCs profiles) can be associated with the effects of cutting, while increasing values are associated with storage time. Decreasing values of PC2 (representing 33.0 % of total variance in VOCs profiles) can also be associated with cutting, and, beyond that, with the distinct effects of the various treatments. Interestingly, sHWT at 55 °C induced the highest alteration of the VOCs profile from that of intact apples (c.f. coloured bars in **Fig. 6.2A**).

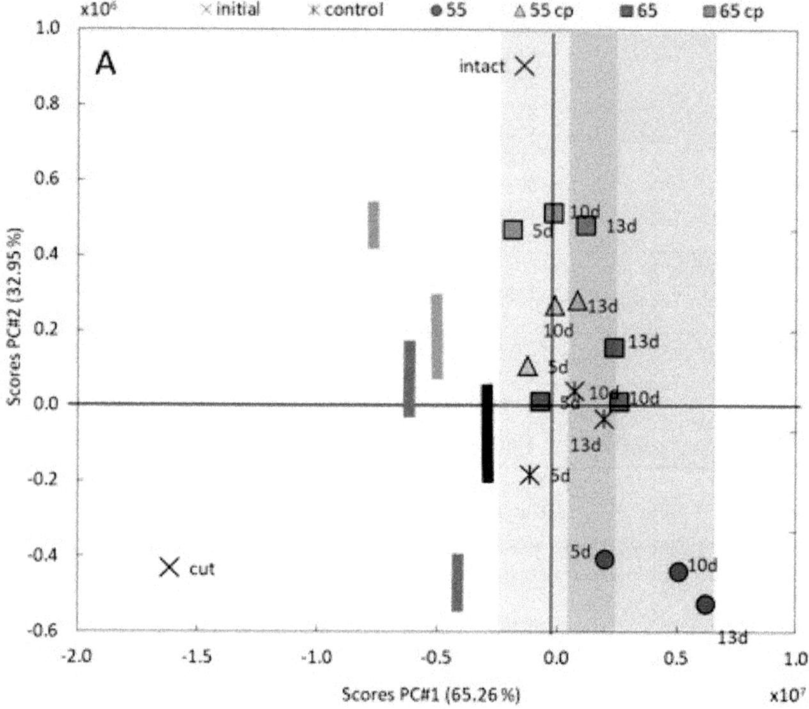

Figure 6.2. (**A**) Scatterplot of principal components analysis (PCA)-results of the variability of VOCs profiles during storage of treated and untreated fresh-cut apple slices. Decreasing scores of PC1 and PC2 can be associated with the effects of cutting; scores values of PC1 are associated with storage time, marked by coloured areas (blue = 3 d and red = 10 d); decreasing scores of PC2 additional associated with distinct effects of the various treatments, marked by coloured bars.

For both PCs, the impact of each single VOC on the total variance in the VOCs profiles can be identified as visualized in **Fig. 6.2B**. Mostly, ethyl acetate (es1) and, to a much smaller extent, ethyl 2-methylbutanoate (es15) and ethanol (al1) are related to effects of the sHWT (PC2), but also to changes during storage (PC1). Cutting-induced changes in the VOCs profiles (both PC decreased) are mainly associated with 2-methylbutyl acetate (es5), but also with isobutyl acetate (es4), butyl acetate (es3) and hexyl acetate (es7).

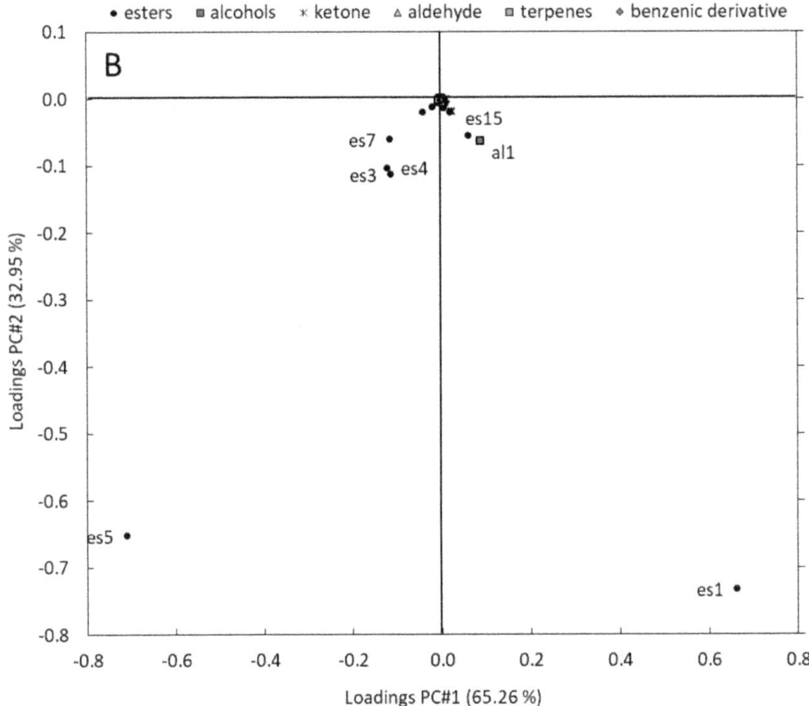

Figure 6.2. (**B**) Loading plot of PCA-results of single-VOC impact on the total variance in VOCs profiles. VOCs with negative loadings of both PCs can be associated with the effects of cutting; VOCs with positive PC1 and negative PC2 loadings are related during storage and distinct effects of the treatments.

3.2. Ethylene Evolution

Cutting immediately and significantly intensified ethylene release of untreated apple slices by approx. 30 %; however, ethylene emission also rapidly, within 4 h, declined to rates below that of intact apples (**Fig. 6.3**). Storage in sugar syrup generally reduced ethylene release as measured after removing the samples from the solution. Compared to controls, sHWT at 55 °C initially (i.e., at day 5 of storage) intensified ethylene emissions of apples slices, while that at 65 °C pronouncedly lowered them. Irrespective of treatments, ethylene emission of all apple slices continuously declined during further storage (**Fig. 6.3**; days 10 and 13).

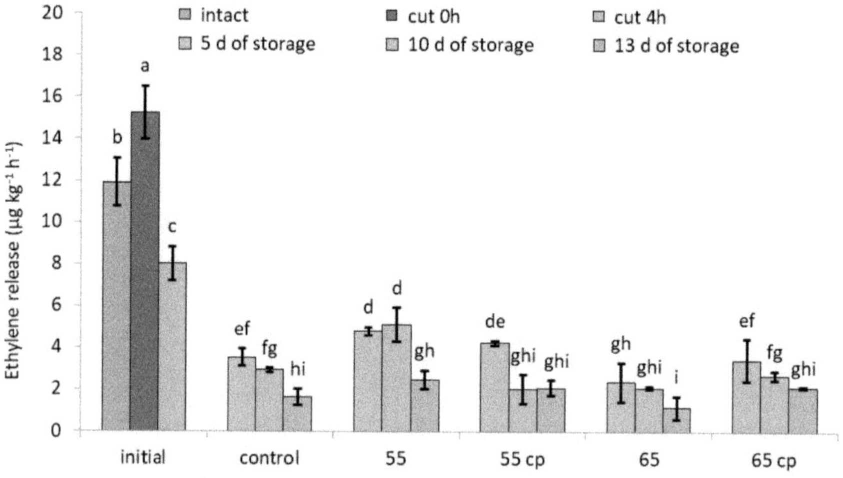

Figure 6.3. Ethylene release rates of short-term hot-water treated and untreated fresh-cut apple slices at days 5, 10 and 13 of storage in sugar syrup at 4 °C, compared to intact apples and untreated fresh-cut apple slices (initial). Given are means ± standard deviation ($n = 3$). Different letters indicate significant differences between means ($p < 0.05$).

3.3. Respiration

CO_2-based respiration rates of untreated 'Braeburn' apples increased 3.4 times immediately after cutting and then slightly declined again within 4 h (**Fig. 6.4**). During storage in sugar syrup, CO_2 release of all samples was higher than that of intact apples. In addition, respiration of sHWT-samples treated at 55 °C without acid treatment was significantly higher than that of the other apple slices at all sampling days.

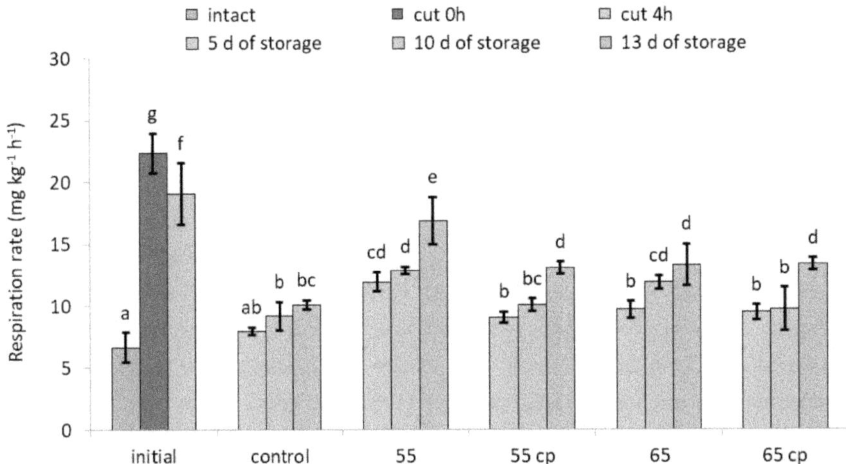

Figure 6.4. Respiration rates of short-term hot-water treated and untreated fresh-cut apple slices at days 5, 10 and 13 of storage in sugar syrup at 4 °C, compared to intact apples and untreated fresh-cut apple slices (initial). Given are means ± standard deviation ($n = 3$). Different letters indicate significant differences between means ($p < 0.05$).

4. Discussion

In the presented experiment, several overlaying and potentially interactive effects were responsible for the development of respective VOCs profiles during storage of apple slices, i.e., cutting, sHW-treatment at two temperatures and storage in sugar syrup with its pronounced effect on O_2 availability. The most pronouncedly emitted VOCs were esters. Similarly, Paillard (1990) reported that esters of acetic, butanoic and hexanoic acids with ethyl, butyl and hexyl alcohols are the most frequent VOCs detected in the headspace of intact apples. The respective composition of the above esters determines the typical fruity (i.e., 'apple-like') cultivar-specific aroma of apples (Fellman et al., 1993). In the presented study, 10 of the identified esters were described as generally important "character impact" VOCs (Dixon and Hewett, 2000). Although the overall composition of the VOCs may clearly vary among different apple cultivars (Poll, 1981), 2-methylbutyl acetate and α-Farnesene are most often the major components in 'Braeburn' apples (Aaby et al., 2002; Baiamonte et al., 2016). This was confirmed by the present results.

4.1. Effect of Cutting on the Release of VOCs, Ethylene and CO_2

The cutting-related increase in acetates probably impacted the aroma of fresh-cut apples, as 2-methylbutyl acetate, butyl-acetate and hexyl acetate are characterized as "character impact compounds" (Dixon and Hewett, 2000). These VOCs are predominantly associated with a characteristic sweet, cox-like apple aroma (Young et al,. 1996). The presented results confirmed earlier findings that cutting inevitably induces physiological wounding-stress responses characterized by temporarily increased ethylene synthesis (Soliva-Fortuny et al., 2002, 2005), respiration activity (Rux et al., 2017) and VOCs release (Bett et al., 2001), which all return to pre-processing levels within 24 h (Finnegan et al., 2013). Emissions of acetate-esters may strongly increase, due to increased lipoxygenase activity (Schreier, 1984) resulting in enhanced esterification of membrane lipids (De Pooter et al., 1983; Rowan et al., 1999). Furthermore, a largely accelerated glycolysis increases the acetyl-CoA availability, and thus also boosts amino-acids production, which serve as precursors to many VOCs (Rowan et al., 1999; Beaulieu, 2006). During storage, however, the concentrations of the aroma-relevant acetates decreased, which may indicate that the cutting induced stress response was reduced or no longer existed. Therefore, the generation of precursors relevant for VOC synthesis was reduced resulting in a decline of emissions. However, the emissions were still higher than those from intact apples. Nevertheless, the typical aroma, appearing immediately after cutting the apples, got continuously lost during storage. This, however, is generally monitored during storage of fresh-cut apples (Beaulieu, 2006).

4.2. Impact of Hot-Water Treatment on the Release of VOCs, Ethylene and CO_2

Even the sHWT at 65 °C did not alter the emission of most VOCs or pronouncedly impacted VOCs profiles but only marginally reduced that of few esters. In contrast to cutting, sHWT did obviously also not affect the synthesis of acetate-esters. This is surprising because a high number of VOCs emanate from the epidermal layers of apples, while only a lower proportion derived from the pulp tissue (Ferreira et al., 2009). However, dipping intact apples for 30 s in hot-water at 55 °C exclusively heated the epidermis and few hypodermal cell layers (Kabelitz et al., 2019). Furthermore, although treatments at higher temperatures (i.e., >55 °C) resulted in faster increase in tissue temperature of deeper cell layers, only 70 °C pronouncedly damaged the epidermis (Kabelitz et al., 2019). Thus, sHWT, in the temperature range used, did not decrease the emission of important "character impact" compounds (Dixon and Hewett, 2000) or increase that of VOCs associated with off-odour. Therefore, sHWT did not reduce the aroma quality. Similarly, sHWT (of up to 65 °C) did not adversely affect colour attributes, tissue strength or other important quality parameters of fresh-cut apple slices (Rux et al., 2019b).

Although the increased ethanol emission may be assumed as critically, the aroma-relevant threshold for this compound is much higher than for other important VOCs (**Table 6.3**). In addition, the ethanol concentrations in total were very low. The observed increase in ethanol emission, therefore, did not negatively affect the aroma quality at all. It may, nevertheless, indicate some sHWT-induced heat stress because ethanol and ethyl-acetate emissions are known to increase in response to and may be used as indicators of this stressor (Forney et al., 2000; Song et al., 2001). Heat stress was reported to alter the glycolytic pathway by disturbing the mitochondrial electron transport, which, similar to reduced O_2 availability, results in increased ethanol formation (Fan et al., 2005). This may also be reflected by the marginally higher respiratory activity observed at the end of storage. Heat treatment-enhanced respiration was also reported earlier for intact 'Granny Smith' and 'Anna' apples (Klein, 1989). In addition, the increased D-limonene emissions of acid-treated sHWT samples may be attributed to major perturbations of the cellular metabolism that result in the expression of multiple genes and, finally, in enhanced terpene emissions (Singh and Johnson-Flanagan, 1998; Pateraki and Kanellis, 2010).

Omitting the post-processing acid treatment increased the ethanol emission, which, however, did not negatively affect the aroma (see above). Hot-water treatments without post-processing acid application only slightly enhanced the alteration of the VOCs profile typical for intact apples in comparison to those with additional acid dipping. Interestingly, sHWT at 55 °C without additional acid treatment clearly intensified the storage-induced physiological responses of fresh-cut apple slices, i.e., it slightly enhanced their respiration activity and ethylene release. This may indicate a direct response of the treated apples to moderate but not to excessive heat or to acid pretreatment (Woodstock and Taylorson, 1981; Charron et al., 1995). Additionally, the emission of ethyl esters is further intensified in sHW-treated (55 °C) apple slices. This is especially obvious for ethyl acetate and ethanol, which, however, both increased in all samples during storage (see Section 4.3). As ethylene may directly regulate the synthesis of VOCs (Bauchot et al., 1998; Beaulieu, 2006; Obando-Ulloa et al., 2009), the increase in ethyl 2-methylpropanoate, ethyl butyrate and ethyl 2-methylbutanoate may be directly related to the enhanced ethylene tissue concentrations. Ethyl acetate correlates with pronounced off-odour (Larsen and Watkins, 1995), while the "character impact compounds" (Dixon and Hewett, 2000) ethyl-butyrate and ethyl 2-

methylbutanoate are generally associated with fruity and apple-like aroma (Rizzolo et al., 1989). The increase in the head space concentrations of the latter compounds, which own aroma thresholds much lower than ethyl acetate (Dixon and Hewett, 2000; Burdock, 2016), may thus even improve the overall aroma of the apple samples (**Table 6.3**).

4.3. Effects of Storage in Sugar Syrup on the Release of VOCs, Ethylene and CO_2

Storing fresh-cut apple slices in sugar syrup pronouncedly affected the development of their VOCs profiles, as indicated by the distinct increase in ethyl ester emissions. Following sHWT, head space concentration of ethyl acetate of samples increased 0.5–1.6 fold compared to controls; it, however, increased approx. 100 times due to the sugar syrup.

The reduction of ethylene and CO_2 release observed after immersion in the syrup, corroborate the findings of Rux et al. (2017). These authors stored apple slices in sugar syrups of different concentrations and attributed major changes in respiration of samples to the reduced O_2 availability due to impaired gas exchange between tissues and ambient air (von Willert et al., 1995; Rux et al., 2017). Therefore, the fast decrease in O_2 and increase in CO_2 concentrations within the tissue increasingly inhibits physiological processes (Rocculi et al., 2006). The increased emissions of ethanol and other alcohols probably indicate semi-anaerobic conditions (Lara et al., 2006).

Likewise, the increased emissions of ethyl esters may be explained by the increased ethanol formation under these conditions. Alcohol acyltransferase (AAT) catalyzes the esterification of carboxylic acids and alcohols (Fellman et al., 1991; Forney et al., 2000), and enhanced ethanol concentrations may enhance the formation of ethyl esters, especially ethyl acetate (Cortellino et al., 2015). In this context, Cortellino et al. (2015) measured a progressive increase of ethyl acetate in fresh-cut apple tissue within 11 d of storage in modified atmosphere, most pronounced at an O_2 concentration of 1 %.

However, in this study, the influence of oxygen availability on VOC synthesis during storage in sugar syrup submerged apple pieces cannot be sufficiently clarified. Further studies focusing on oxygen availability are necessary.

5. Conclusions

In a semi-practical approach, this study comprehensively evaluated the potential food quality-related effects of sHWT (at 55 °C and 65 °C), cutting and/or post-processing acid treatment and their interactions on the development of aroma-relevant VOCs as well as on the physiological indicators ethylene synthesis and respiration activity in fresh-cut apple slices stored in sugar syrup. Cutting temporarily enhanced ethylene synthesis, respiration activity and VOCs production. In this context, 2-methylbutyl acetate, butyl acetate and hexyl acetate were identified as most relevant cutting-enhanced VOCs, but their emissions declined again during syrup-storage. Syrup-storage increased ethanol and ethyl-esters, and in particular ethyl-acetate, which might be attributed to reduced O_2 availability under this condition. Solely applying sHWT at both 55 °C and 65 °C enhanced the respiration of apples slices, which potentially indicated marginal heat stress. This, however, only slightly intensified ethylene release but increased ethanol emission of samples, indicating alterations in respiratory pathways. Additionally, sHWT at 65 °C resulted only in minor declines of few esters. Nevertheless, the general impact of sHWT on VOCs profiles of apples slices was low, especially in comparison to storage in sugar syrup and did not negatively affect aroma. The omission of pre-processing acid application did not negatively affect the aroma to a crucial degree. Consequently, sHWT might replace the post-processing acid treatments for fresh-cut products.

7 General discussion

In the following chapter 7, important aspects of fresh-cut apple metabolism and investigated sanitation methods are summarized. In addition, the results of the different investigations are comprehensively discussed.

7.1 Impact of processing and storage on physiological parameters

Treatments, such as cutting, application of organic acid solutions or hot water treatments, and storage conditions, such as modified atmosphere (MA) or sugar syrup immersions, induce physiological responses of fresh-cut apples, i.e., indicated by changes in VOC emissions, respiration and ethylene synthesis. All of the above processes and conditions can affect the respiration, and the synthesis and the release of ethylene and VOCs (El Hadi et al., 2013). Especially the biosynthesis of VOCs is pronouncedly affected by various interactions with respiration and ethylene synthesis (Yabumoto et al., 1977; Sanz et al., 1997; Fellman et al., 2000; Espino-Díaz et al., 2016). The knowledge on the impact of the O_2 availability on the physiological responses in fresh-cuts is still limited. This thesis comprehensively evaluates these potentially synergistic and interactive effects, focusing on the emission of VOCs. Several overlapping effects were responsible for the development of respective VOCs profiles during storage of apple slices.

The cutting of apples induced wound responses resulting in a temporarily accelerated metabolism indicated by the enhanced ethylene release and respiration activity, and by the strongly increased ester production, especially that of acetate-esters. These results confirmed earlier findings (Bett et al., 2001; Soliva-Fortuny et al., 2002, 2005; Mahajan et al., 2014). The pronounced increase in the emissions of acetate-esters was most probably due to the accelerated glycolysis and the increased lipoxygenase activity. In this context, 2-methylbutyl acetate, butyl acetate and hexyl acetate were identified as most relevant cutting-enhanced VOCs. The aroma of these compounds is e.g. described as sweet fruity and ripe, as well as associated with characteristic apple aroma (Dixon and Hewett, 2000), and can therefore be considered positive as they create the typical intensification of apple aroma after cutting. Mainly the synthesis and the emission of 2-methylbutyl acetate were intensified, a compound primarily produced from amino acids degradation. This effect could probably be attributed to the boosted glycolysis, which increases the acetyl-CoA availability and, thus, also the amino-acids production (Rowan et al., 1999; Beaulieu, 2006). The shift from β-oxidation to lipoxygenase activity after the disruption of biomembranes due to fruit tissue damages may also provide membrane lipids for enhanced ester production (De Pooter et al., 1983; Rowan et al., 1999). In addition, cutting also activated other enzymatic systems, which accelerated degradation of injured cell membranes (Nicoli et al., 1994; Toivonen and DeEll, 2002). During storage, however, the ethylene release and the concentrations of acetates diminished. This finding nicely reflects other reports that the enhanced activity of the above mentioned metabolic pathways is only temporary, returning to the normal level within 24 h (Finnegan et al., 2013). In addition, the rate of the decline in the acetate ester emissions (except that of ethyl acetate) depended on the O_2 availability, e.g. if fresh-cut apples immersed in sugar syrup or not were compared.

During storage in either ambient air or under modified atmosphere, emissions of ethanol and ethyl esters, especially of ethyl acetate, were increased. Increased emissions of both, ethanol and ethyl acetate are related to the off-odour of apples and considered negative (Wright et al., 2015). The ethanol emissions may be enhanced by the membrane damage, which disturbs respiration and intensifies the accumulation of pyruvate, the precursor of ethanol (Ke et al., 1994). At O_2 concentrations close to 0 %, the ethanol emissions were further enhanced. This might be due to anaeropiosis, which also results in the accumulation of pyruvate. The esterification by alcohol acetyl-CoA transferase (AAT) again led to the formation of ethyl esters (Forney et al., 2000). This synthesis of ethyl esters consumes high amounts of acetyl-CoA, and thereby decreases the concentrations of other acetate esters due to competitive reaction (Kollmannsberger and Berger, 1992; Ke et al., 1994). The emissions of butyl acetate, 2-methylbutyl acetate, hexyl acetate, ethyl 2-methylpropanoate and ethyl butyrate, characteristic for the apple aroma, were significantly reduced at O_2 concentrations close to 0 %. This reflects a low ethylene synthesis during storage under these conditions, because those compounds are directly regulated by ethylene (Beaulieu, 2006; Schaffer et al., 2007; Obando-Ulloa et al., 2009).

Immersion of fresh-cut apples in sugar syrup and pure orange juice created "modified atmospheres" inside the tissues (diminished O_2 availability) during storage, which directly retarded the physiological processes. Irrespective of the medium used, immersion of apple slices in liquids strongly reduced CO_2 release. This was due to the additional gas diffusion barrier and, hence, increased resistance to gas diffusion, generally restricting the gas exchange between the tissues and the ambient air (von Willert et al., 1995). The respiration activity and, therefore, the CO_2 release of apple slices depended on kind of liquid or the sugar concentrations used (**7.2.1**). However, immersion generally increased the risk of anaerobic respiration resulting in synthesis of fermentative metabolites, e.g. ethanol.

In this context, the presented findings revealed two distinct responses of the VOCs emissions during storage of apple slices immersed in 20 % sugar solution. In one study (Chapter 4), immersion led to the continuous decline of ethyl esters emissions, while the emissions in another investigation (Chapter 6) strongly increased to a multiple of the initial rates (**Tabelle 7.1**). The decrease of ethyl esters emission of submerged slices in the first study was in contrast to increased emissions of those stored in normal air or under MA. This indicated metabolic changes, which were independent of O_2 availability as sole factor. The strong increases in ethyl ester emissions in the second investigation were attributed to reduced O_2 availability under this condition. However, the storage conditions were very similar and characterized by very low O_2 availability in both studies. This was revealed by the increased ethanol emissions, which indicated semi-anaerobic conditions (Lara et al., 2006). The direct comparison of both investigations revealed an interesting feature.

The initial ethanol emissions were in a similar range (7.3 - 9.7 nL L^{-1}) and declined to 2.2 nL L^{-1} after cutting in both investigations. While the emissions in the second investigation only increased to 13.1 nL L⁻1 after 10 d of storage, they were about four times as high in the first investigation (51.1 nL L^{-1}). Alcohol acyltransferase (AAT) catalyses the esterification of alcohols (Fellman et al., 1991; Forney et al., 2000) to ethyl esters, and therefore, enhanced ethanol concentrations may also enhanced the formation of ethyl esters, especially that of ethyl acetate (Cortellino et al., 2015). This was only observed in the second investigation, where ethyl esters emission strongly increased, while that of ethanol remained at rates similar to that before cutting. In the first investigation, only ethanol emissions increased, while that of ethyl ester continuously declined. Therefore, it can be assumed that the esterification is inhibited or prevented resulting in an accumulation of ethanol. Thereby, the esterification depended on the availability of alcohol and of acetyl-CoA or acyl-CoA (Paillard, 1979; De Pooter et al., 1981; Knee and Hatfield, 1981; Fellman et al., 1991). Differences in the ester profiles, therefore, could be attributed to availability of precursors. However, reduced ethyl ester emissions may indicate either a lack of CoA as important precursor or a reduced AAT activity. Esterification by AAT may be partially inhibited by the lowered pH due to the high CO_2 concentrations accumulating in the fruit tissue during immersion (Ke et al., 1994).

Table 7.1. Semi-quantitative concentrations (nL L^{-1}) of VOCs, emitted by intact, freshly cut (cut) and fresh-cut apple slices on day 10 of storage in sugar syrup at 4 °C from to different investigations (controls shown in Paper II and IV). VOCs concentrations given resulted from the emissions of 20 apple slices hermetically enclosed in a 1 L-glass jar at 4 °C for 1 h. Given are means ($n = 3$). Different letters indicate significant differences between means ($p < 0.05$). The different colours should help to indicate increasing (green) or decreasing (red) emissions of VOCs compared to intact apples (yellow).

VOC	Chapter 3 (O_2 availability)			Chapter 6 (sHWT)		
	intact	cut	10 d	intact	cut	10 d
Cumulative VOC concentration	246 a	1846 b	147 a	87 a	618 c	364 b
Ethyl acetate	15.9 a	47.9 b	21.4 a	0.86 a	1.98 a	104 b
Propyl acetate	3.06 a	24.8 b	2.02 a	nd	2.91 a	3.35 a
Butyl acetate *	29.3 b	307 c	0.98 a	2.53 a	56.0 c	20.0 b
Isobutyl acetate	3.59 a	52.7 b	1.56 a	1.52 a	64.1 b	16.7 a
2-methylbutyl acetate *	35.9 a	621 b	6.88 a	25.2 a	323 c	94.1 b
Pentyl acetate *	3.59 a	43.1 b	nd	0.78 a	13.1 c	2.64 a
Hexyl acetate *	40.0 a	374 b	1.09 a	3.15 a	59.8 b	3.48 a
2-Hexen-1-yl acetate	0.30 a	22.3 b	0.08 a	nd	22.9 b	nd
Ethyl propionate	4.63 a	12.1 b	0.87 a	1.46 a	nd	3.27 b
Ethyl 2-methylpropanoate	0.69 a	2.34 a	1.57 a	nd	nd	0.97
Butyl propionate*	2.59 b	15.0 c	0.90 a	nd	nd	nd
Hexyl propionate*	2.76 a	2.72 a	nd	nd	nd	nd
Methyl butyrate	2.53 a	11.5 b	0.24 a	0.63 a	4.20 b	1.13 a
Methyl 2-methylbutyrate *	1.19 a	8.04 b	1.00 a	nd	nd	1.97
Ethyl butyrate *	41.8 a	152 b	16.4 a	nd	1.64 a	10.9 b
Ethyl 2-butenoate	nd	nd	0.24	nd	nd	0.70
Ethyl 2-methylbutanoate *	12.4 a	63.2 b	16.9 a	nd	0.98 a	25.1 b-c
Butyl butyrate*	6.85 b	14.0 a	nd	nd	nd	nd
Hexyl butyrate *	8.67 a	6.66 b	0.79 c	5.24 a	5.39 a	1.00 b
Hexyl 2-methylbutanoate *	nd	nd	nd	5.56 a	8.20 b	11.2 c
Ethyl valerate	0.03 a	0.69 b	nd	nd	nd	0.34
Methyl hexanoate	0.53 a	2.03 a	0.88 a	nd	2.09 b	0.41 a
Ethyl hexanoate	4.06 a	20.0 a	2.38 a	nd	1.56 a	7.15 b
Hexyl hexanoate *	3.85 a	2.37 b	1.41 b	2.11 a	1.92 a	0.71 b
Ethanol	7.27 a	2.19 a	51.1 b	9.72 b	2.21 a	13.1 c
2-(2-Ethoxyethoxy)ethanol	0.57 a	0.56 a	0.60 a	0.65 a	0.62 a	0.30 b
2-Methyl-1-propanol	0.70 a	1.66 c	1.06 b	0.86 a	2.99 b	0.92 a
1-Butanol *	nd	nd	nd	1.51 a	2.82 b	1.59 a
2-Methyl-1-butanol	3.68 a	9.46 b	4.56 a	3.24 a	9.02 b	4.63 a
1-Hexanol *	nd	nd	nd	1.44 a	3.94 b	3.06 b
2-Butanone	0.41 a	0.08 a	0.57 a	2.67 a	0.33 a	0.73 a
1-Penten-3-one	nd	nd	nd	nd	nd	3.71
Hexanal *	0.55 a	4.45 b	0.44 a	0.50 a	1.92 a	4.49 b
Estragole	0.66 a-b	0.75 b	0.45 a	0.30 a	0.41 a-b	0.52 b
D-Limonene	0.64 a	1.23 a	1.31 a	3.27 a	4.93 a	4.74 a
α-Farnesene	7.03 a	13.4 b	9.37 a	13.8 a	19.1 a	17.4 a

* marked VOCs described as generally important "character impact" compounds in apples (Dixon and Hewett 2000). nd = not detected/below detection limit; dark green = ≥500 % increase; light green = ≥33 % increase; yellow = ≤33 % increase/decrease ; light red = ≥80 % decrease; dark red = ≥500 % decrease.

The transfer of fresh-cut apple slices from MA conditions or syrup immersion to normal air at the end of the storage led to a fast change in metabolism due to the changed O_2 availability. Low O_2 concentrations during storage enhanced the ethylene synthesis after exposure of slices to normal air. Respiration adapted rapidly to normal air conditions and only slightly depended on these. CO_2 release generally increased after taking samples out of the liquids or the MA. This mainly resulted from the pronouncedly lower diffusion resistance in the air than in the liquid.

7.2 Evaluation of the preservation methods

7.2.1 Effect of immersion of fresh-cut apple slices

Postharvest storage of fresh-cut fruit (e.g. apples) in sugar-syrup or suitable fruit juice solutions are practically used in practise to extend product shelf life by preventing enzymatic and oxidative browning and water losses, and to retard respiration and ethylene syntheses and other metabolic processes. This study comprehensively evaluates these practices because scientific information on the physiological responses and changes in the quality of fresh-cut apples immersed in sugar solution or fruit juice are lacking.

Both sugar syrup and pure orange juice were shown to prevent browning effects and considerably reduce respiration activity of immersed fresh-cut 'Elstar' and 'Braeburn' apple slices to a certain extent. These effects are mainly due to the reduced oxygen concentration in fruit tissue after immersion in the liquid media, which represent a pronounced gas diffusion barrier and, thus, reduce the gas exchange from the ambient to the tissues and vice versa (von Willert et al., 1995). Diminished O_2 availability in the apple tissue retards the respiration activity and reduces the oxidative browning analogue to the modified atmosphere conditions. Furthermore, adding browning inhibitors and/or ascorbic/citric acids to the sugar solutions, as well as the natural acidity of orange juice also prevents browning of apple slices due to the lowering of the pH, finally reducing the PPO activity (Weemaes et al., 1998). However, the immersion of fresh-cut apples slices also increases the risk of anaerobic respiration (see Chapter **7.1.**) and could not prevent or minimise the microbiological growth.

The use of orange juice as a preservation medium not only increased the microbial loads of the apple slices, but also promoted the visible growth of moulds on the medium surface. Considering both, the immersed apples and the used medium, orange juice appears unsuitable for the storage of apple slices over several days in terms of food safety. Furthermore, any dilution of the orange juice clearly further enhanced quality losses of samples. It increased the metabolic activity, the browning effects and the microbial growth. It also reduced the tissue strength and gradually declined the soluble solid and acidity contents of the immersed apple sliced compare to pure orange juice depending on the dilution factor. In contrast, short-term (up to 2 d) immersion of fresh-cut apple slices in pure orange juice is of particular interest for the gastronomic self-service (catering, hotel, canteen etc.), because this immersion effectively prevents browning and water loss.

For storage of fruit slices in sugar syrups, sugar contents of 13 – 20 % were optimal and resulted in the least browning and a minimal respiration activity. Reduction of sugar concentration of the syrup below 13.4 %, which reflects the total soluble solids content of investigated apple slices, negatively affected the sample quality during storage. It increased metabolic activity of stored apple slices, declined their soluble solid content and reduced tissue strength. These negative effects were mainly due to the low osmolarity of the medium, which resulted in hyper-turgidity of slices, led to tissue lysis and disintegration of mature apple cells. Furthermore, lower sugar concentrations enhanced the growth of aerobic mesophilic bacteria. Nevertheless, also the increase in the sugar content up to 30 % induced adverse metabolic changes and quality losses. Hyper-osmotic solutions led to the dehydration of the immersed apple slices and, thus, mass losses and low tissue strength. Therefore, sugar syrups with sugar contents in the range of 13.4 – 20 % turned out to be optimal for the maintenance of quality and safety of stored fresh-cut fruit.

However, even at a sugar concentration the syrup of 20 %, the impact of immersion on a few quality parameters varied with the apple cultivars, and the storage temperature and time used (**Table 7.2**). In 'Elstar' apples, storage at 13 °C up to 6 d increased the tissue strength of fresh-cut slices (I), while storage at 4 °C for 10 - 13 d resulted in its decline in 'Braeburn' apple slices (II; III). This cultivar-depending (Kim et al., 1993) softening may be attributed to acid-related degradation of pectins (Zhu et al., 2007) and to the stage of fruit maturity (Gorny et al., 2000; Rojas-Graü et al., 2007). Therefore, storage in sugar syrup may be restricted to apple cultivars that are prone to fast tissue browning but retain firmness during shelf life.

Soluble solid contents were retained almost at the initial quality but increased, decreased or remained constant in the different experiments. These slight differences in SCC probably also cultivar-specific and depended on the stage of fruit maturity. The changes in total acidity were also not uniformly and, as thus, difficult to evaluate. However, desired final acid contents can be easily obtained by adding respective amounts of ascorbic acid to the sugar solution.

Table 7.2. Comparison of selected quality parameters (SSC: total soluble solids, TA: total acidity; cumulative VOC concentrations; microbial load for TAMB: total aerobic mesophilic bacteria, yeasts and moulds) of fresh-cut apple slices stored in 20 % sugar syrup, measured in the different investigations (I: Chapter 3; II: Chapter 4; III; Chapter 5 & 6). Shown are partially the initial values and the corresponding values on day 6 (I) or day 10 (II & III) of storage.

Investigation	**I**		**II**		**III**	
Cultivar	'Elstar'		'Braeburn'		'Braeburn'	
Temperature (°C)	13		4		4	
Max. storage days	6		10		13	
Day of storage	0	6	0	10	0	10
Tissue strength (N)	2.0	2.2	8.7	3.9	7.4	3.2
SSC (% Brix)	13.4	14.2	12.0	11.1	12.5	12.2^{ns}
TA (mg kg^{-1})	3.5	2.8	5.1	4.8	4	6
Cumulative VOC conc. (nL L^{-1})	-		246	147	87	364
TAMB/Yeast/Mould (log CFU g^{-1})	5/5/4		4/6/0		4/6/1	

^{ns} not significant

The immersion of apple slices in sugar syrups clearly reduced the emissions of VOCs compared to atmospheric storage due to enhanced gas diffusion resistance in liquid. However, semi-anaerobic conditions could increase the synthesis of undesirable ethanol and ethyl esters. It was shown that immersion of apple slices in syrup limited or disturbed the esterification of ethyl esters from ethanol (see Chapter **7.1**), resulting in the continuous accumulation of the latter and the decline of ethyl esters emission. The reduction of VOCs emissions led to a pronounced loss of aroma (cutting related acetates) but also prevented off-odour (ethanol or ethyl acetate). Therefore, the use of sugar syrup for storage of apple slices may extend their shelf life, though at the expense of a lower aroma quality.

However, immersion also enhanced microbial growth, especially that of yeasts, which may be due to the increased availability of water and sugars. At a storage temperature of 13 °C, yeasts and moulds exceeded the limits defined by the German Association for Hygiene and Microbiology (DGHM, 2011) for fresh-cut and packaged fruits after 6 d of storage. At 4 °C, only yeast counts exceeded these limits, while moulds did not cause any problem probably due to the low pH combined with the high sugar contents in the syrup as 4 °C, which effectively suppressed their growth (Tapia de Daza et al., 1996). Yeast counts in the syrup also exceeded the accepted limits after 10 d of storage. Thereby, microbial quality of the syrup also plays a major role because the sugar solutions present a large proportion of the total mass of stored fresh-cut fruit salads. It was shown that syrups may be contaminated by microorganisms, adherent to the slices. Particularly yeasts grew faster in the syrup than on apple slices. Therefore, in the present study, yeasts are the most critical parameter for the shelf life of fresh-cut apples immersed in sugar syrup. This highlighted the great importance of the initial hygienic state of the apples to be processed.

In the present study, the impact of the liquid media was examined for the first time. This is especially important for sugar solutions, which are already used in practice. The presented results highlighted the advantages but also the limitations of these storage methods. Orange juice should be primarily used for periods of less than two days due to its low microbial stability. In contrast, syrups with sugar concentration of 20 % guaranteed a shelf life of at least 10 d. For the latter, it is recommended to use apple cultivars, which are prone to rapid tissue browning, but are not susceptible to softening. Relevant additives to the sugar syrup (e.g. ascorbic or citric acid) may specifically regulate the acidity and the vitamin C content. The growth of yeasts was identified as a critical parameter, which emphasizes the importance of adequate sanitation before processing and storage. Shelf life of fresh-cut apple slices can thus be well extended if processing and storage is combined with suitable sanitation methods.

7.2.2 Potential effects of short-term hot-water treatments

Microbial spoilage is a major factor limiting the shelf life of apple slices immersed in sugar syrup and the removal or inactivation of microbes is an essential prerequisite of maintenance of quality and safety of fresh-cut fruit. Since several decades, hot-water treatment has been identified as a gentle, sustainable and environmentally conscious sanitation technique, which is also suitable for organic production (Maxin et al., 2014). In the context of fresh-cut products, especially short-term hot-water treatments (sHWT) may effectively reduce microbial contaminations to guarantee a commercially required shelf life of 10 d. This study investigated the effect of sHWT on quality, fruit aroma and physiology of immersed fresh-cut 'Braeburn' apple slices. Furthermore, the optimal conditions for sustainable processing were evaluated. Finally, it was analysed whether sHWT could potentially supplement or replace post-processing chemical treatments in the production of organic fresh-cut fruit salads.

In terms of product safety aspects, sHWT significantly reduced the initial microbial load of apples similarly to the common post-cutting acid treatment. However, sHWT could not extend the shelf life of apple slices, even though results showed a temporary inhibition on TAMB and yeast growth during the initial 5 d of storage in syrup. In the present study, yeasts are most relevant for the maximum obtainable shelf life, since only yeast counts exceeded the limits (5 log CFU g^{-1}) defined by the DGHM (2011) after 10 d of storage. The increase of the treatment temperature from 55 °C to 65 °C only marginally improved TAMB reduction. However, the combination of post-cutting acid treatment and the sHWT did not or only minimally enhance the reduction of microbes.

The results support the conclusion that direct heat inactivation of microorganisms is not the main mechanism of HWT sanitation (Kabelitz et al., 2019), as the temperature rises from 55 °C to 65 °C has no effect. Rather, the study supports the assumption that the temporary melting of the epidermal wax layer, which produces a physical barrier to subsequent infections, may play an important role. This is derived from the circumstance that a temperature increase to 65 °C does indeed lead to a significant growth reduction of microbes for uncut apples. This cannot be expected with fresh-cut apples as the protective function of the wax layer is eliminated during cutting.

Pre-processing sHWT at both 55 °C and 65 °C did not adversely affect important quality parameters of fresh-cut apple slices. In addition, the combination of sHWT and the post-cutting acid treatment did not further improve quality maintenance. In detail, sHWT did not induce any peel or tissue browning. Apple slices treated at 65 °C, however, experienced a slight reduction of the red peel colour due to the degradation of anthocyanins, while the yellow colour increased compared to controls. Results of the present study clearly highlighted that sHWT did not affect the sugar contents, indicated by SSC, the total titratable acid or the vitamin C contents of apple slices. Although tissue strength continuously declined during storage in sugar syrup, sHWT tended to diminish softening, particularly in those samples that were not additionally treated with acid solutions. In the used temperature range, sHWT did not decrease the emission of important "character impact" VOCs (Dixon and Hewett, 2000) or increase that of VOCs associated with off-odour. This shows that heat has only a minor impact on VOCs synthesis, since the majority of these compounds are emanating from the epidermal layers (Ferreira et al., 2009), the location of the highest heat impact. The omission of the pre-processing acid application did also not negatively affect the aroma to a crucial degree. In terms of product quality, similar studies on sHW-treated (up to 65 °C for 60 s) intact 'Braeburn' apples also highlighted the lack of negative effects (Kabelitz et al., 2019; Herppich et al., 2020). Beyond that, Herppich et al. (2020) did not observe any effects of sHWT on chlorophyll or anthocyanin contents of treated apples.

In contrast to quality parameters, solely applying sHWT at both temperatures enhanced the metabolic activity (as indicated by respiration) of apple slices, which potentially reflects marginal heat stress. Ethylene release and ethanol emission were slightly intensified due to sHWT, pointing to alterations in respiratory pathways.

Ethanol emissions are known to increase in response to heat stress, which affects the respiratory pathways by disturbing the mitochondrial electron transport, similarly to the reduced O_2 availability (Forney et al., 2000; Song et al., 2001; Fan et al., 2005). As the sensorial threshold of ethanol is much higher than that of other important VOCs and its effective concentration was very low, this increase did not negatively affect the aroma quality. In addition, sHWT combined with the acid treatment raised the emissions of D-limonene, which may result from the enhanced expression of multiple genes as a response of the major perturbations of the cellular metabolism (Singh and Johnson-Flanagan, 1998; Pateraki and Kanellis, 2010). However, the increase in D-limonene can probably be considered unproblematic or even positive, as it produces a fruity and grassy aroma (Zepka et al., 2014).Increasing the treatment temperature from 55 °C to 65 °C marginally lowered the emission of few esters. In contrast, omitting the post-processing acid treatment slightly altered the VOCs profile, but further increased the ethanol emissions. Interestingly, sHWT at 55 °C without additional acid treatment clearly intensified the storage-induced physiological responses of fresh-cut apple slices. These responses include the increase of respiration, ethylene release and emission of ethanol and ethyl esters, especially ethyl acetate. Therefore, the higher HW-treatment temperature of 65 °C seems to be gentler, as the mentioned reactions are assessed negatively. However, a detailed investigation of this relationship appears to be necessary.

The photosynthetic apparatus was referred to as the major initial target of heat stress in apple skin cells (Schlüter et al., 2009; Baier et al., 2013) and, therefore, was used as a sensitive marker for heat stress effects (Herppich et al., 1997; Matyssek and Herppich, 2018). Herppich et al. (2020) described a sHWT-induced inhibition of photosynthetic performance in intact apples under comparable treatment conditions. In this study, changes in chlorophyll fluorescence imaging (CFI) parameters and spectral factors were analysed in fruit of two coloured ('Braeburn' and 'Fuji') and two green-ripe apple cultivars ('Greenstar' and 'Granny Smith'), sHW-treated in the range of 44 °C to 70 °C for 0.5 min to 5 min. The green-ripe cultivars showed less sensitivity to sHWT compared to red-coloured cultivars. Herppich et al. (2020) addressed the stage of fruit development as an important determinant of heat stress sensitivity (Fan et al., 2011) and mentioned a potential influence of fruit peel colour on the sensitivity to HWT effects. Furthermore, the thresholds of heat inhibition observed for 'Braeburn' cultivar were distinctly lower than those of 'Fuji' (Herppich et al., 2020). These highly specific and varied physiological responses of apples of different cultivars to heat treatments were also reported by other authors (e.g. Kim et al., 1993; Shao et al., 2007; Fan et al., 2011; Hengari et al., 2016; Schloffer and Linhard, 2016).

For the first time, the effects of sHWT and/or of post-processing acid treatment both combined with sugar syrup-storage on fruit quality and shelf life were comprehensively and comparatively evaluated for fresh-cut apple slices. Despite the heat-stress induced physiological responses, sHWT up to 65 °C for up to 1 min has no negative impact on fruit quality. Summarizing, the presented results pointed out that the use of sHWT eliminates the need for post-processing acid treatment. Consequently, sHWT might replace the post-processing acid treatments for fresh-cut apple products. Application of sHWT can lower the use of chemicals and may reduce production costs, although the energy expenditure must be considered.

8 Conclusion

Object of this thesis was the evaluation of complete immersion of fresh-cut apples in sugar solution or orange juice on characteristic quality attributes and product physiology. Furthermore, this thesis investigated the potential synergistic interactive effects caused by the sugar syrup immersion of apple slices and the O_2 availability on VOCs emissions. In addition, the potential implications of short-term hot-water treatment (sHWT) in the temperature range of 55 – 65 °C on product quality, fruit aroma related volatiles and physiology of immersed fresh-cut 'Braeburn' apple slices were investigated. This includes the ability to reduce effectively microbial contaminations and evaluation of possible positive or negative effects on product quality.

Following the object of this thesis to evaluate the effects of complete immersion of fresh-cut apples in sugar solution or orange juice on quality and physiology, the obtained results point out that both treatments prevent tissue browning and considerably reduce respiration rates both in 'Elstar' and 'Braeburn' apple slices. Irrespective of the medium used, immersion of fruit slices created "modified atmospheres" inside the tissues (diminished O_2 availability), which directly retarded physiological processes such as respiration. However, immersion also increased the risk of anaerobic respiration resulting in synthesis of fermentative metabolites, e.g. ethanol and ethyl esters. Short-term (up to 2 d) immersion of fresh-cut apple slices in pure orange juice to prevent browning and water losses is of particular interest for the gastronomic self-service (catering, hotel, canteen etc.). However, the use of orange juice as a preservation medium appears unsuitable for the safe storage of apple slices over several days. Furthermore, any dilution of the orange juice clearly further increased the metabolic activity, the browning effects and the microbial growth. It also reduced the tissue strength and gradually declined the soluble solid and acidity contents of the immersed apple sliced depending on the dilution factor.

For storage of apple slices in sugar syrups, sugar contents of 13 – 20 % were optimal and resulted in the least browning and a minimal respiration. Reduction of sugar concentration increased metabolic activity of stored apple slices, lowered their soluble solid content, reduced tissue strength and enhanced the growth of aerobic mesophilic bacteria. The increase in the sugar content up to 30 % led to the dehydration of the immersed apple slices and, thus, mass losses and low tissue strength.

The immersion in 20 % sugar solution clearly reduced the emissions of VOCs compared to atmospheric storage. The reduction of VOCs emissions probably led to a pronounced loss of aroma but also prevented off-odour. In this context, immersion can induce metabolic changes, which are independent of O_2 availability as sole factor. It was assumed that the esterification by AAT is inhibited or prevented resulting in an accumulation of ethanol. The impact of immersion on tissue strength varied with the apple cultivars and stage of maturity. Therefore, storage in sugar syrup may be restricted to apple cultivars that are prone to fast tissue browning but retain firmness during shelf life. Relevant additives to the sugar syrup (e.g. ascorbic or citric acid) may specifically regulate the acidity and the vitamin C content.

However, immersion also enhanced microbial growth, especially that of yeasts. Nevertheless, syrups with sugar concentration of 20 % guaranteed a shelf life of at least 10 d if fruit slices are stored at 4 °C. The growth of yeasts was identified as a critical parameter, which emphasizes the importance of adequate sanitation before processing and storage. The microbial quality of the syrup also plays a major role because of their large proportion of the total mass. It was shown that syrups may be contaminated by microorganisms, adherent to the fruit slices. Particularly yeasts grew faster in the syrup than on apple slices. This highlighted the great importance of the initial hygienic state of the apples to be processed.

Pre-processing sHWT at both 55 °C and 65 °C significantly reduced the initial microbial load of apples but could not extend the shelf life of apple slices. However, the total aerobic mesophilic bacteria (TAMB) and yeast growth were temporary inhibited during the initial 5 d of storage. The increase of the treatment temperature from 55 °C to 65 °C only marginally improved TAMB reduction. Nevertheless, sHWT did not adversely affect important quality parameters of fresh-cut apple slices and did not decrease the emission of important "character impact" VOCs or increase that of VOCs associated with off-odour. However, solely applying sHWT enhanced the metabolic activity, e.g. the respiration of apple slices, which potentially reflects marginal heat stress at both temperatures and ethylene and ethanol emission, the latter pointing to alterations in the respiratory pathways. The combination of post-cutting acid treatment and the sHWT did not or only minimally enhance the reduction of microbes and did not further improve quality maintenance. Therefore, sHWT might replace the post-processing acid treatment for fresh-cut products to lower the use of chemicals and may, thus, reduce the production costs.

References

Aaby, K., Haffner, K., & Skrede, G. (2002). Aroma quality of Gravenstein apples influenced by regular and controlled atmosphere storage. *LWT-Food Sci. Technol.*, 35(3), 254-259. https://doi.org/10.1006/fstl.2001.0852

Abadias, M., Canamas, T.P., Asensio, A., Anguera, M., & Vinas, I. (2006). Microbial quality of commercial 'golden delicious' apples throughout production and shelf-life in Lleida (Catalonia, Spain). *Int. J. Food Microbiol.*, 108, 404–409. https://doi.org/10.1016/j.ijfoodmicro.2005.12.011

Abbott J.A., & Harker, F.R. (2004). Texture. In *The Commercial Storage of Fruits, Vegetables, and Florist and Nursery Stocks, USDA Handbook No. 66, 3rd edition*. Gross, K.C., Wang, C.Y., & Saltveit, M.E. (Eds.) Agricultural Research Service: Washington, DC.

Ahvenainen, J.M. (1996). New approaches in improving the shelf life of minimally processed fruits and vegetables. *Trends Food Sci. Technol.*, 7, 179–186. https://doi.org/10.1016/0924-2244(96)10022-4

Ahvenainen, R. (2000). Minimal processing of fresh produce. In *Minimally Processed Fruits and Vegetables*. Alzamora, S.M., Lopez-Malo, A., Tapia, M.S. (Eds.), Aspen Publisher Inc.: Gaithersburg, MD, USA.

Al-Ati, T., & Hotchkiss, J.H. (2003). The role of packaging film permselectivity in modified atmosphere packaging. *J. Agric. Food Chem.*, 51, 4133–4138. https://doi.org/10.1021/jf034191b

Alzamora, S.M., Tapia, M.S., Argaíz, A., & Welli, J. (1993). Application of combined methods technology in minimally processed fruits. *Food Res. Int.*, 26, 125-130. https://doi.org/10.1016/0963-9969(93)90068-T

Ardö, Y. (2006). Flavour formation by amino acid catabolism. *Biotechnol. Adv.*, 24(2), 238-242. https://doi.org/10.1016/j.biotechadv.2005.11.005

Artés, F., Gómez, P.A., & Artés-Hernández, F. (2006). Modified atmosphere packaging of fruits and vegetables. *Stewart Postharvest Rev.*, 2(5), 1-13. https://doi.org/10.2212/spr.2006.5.2

Artés, F., Gómez, P.A., & Artés-Hernández, F. (2007). Physical, physiological and microbial deterioration of minimally fresh processed fruits and vegetables. *Food Sci. Technol. Int.*, *13*(3), 177-188. https://doi.org/10.1177/1082013207079610

ASABE (2008). Compression test of food materials of convex shape. ASABE Standards, ASAE S368.4 DEC2000 (R2008).

Austin, J.W., Dodds, K.L., Blanchfield, B., & Farber, J.M. (1998). Growth and toxin production by *Clostridium botulinum* on inoculated fresh-cut packaged vegetables. *J. Food Prot.*, 61(3), 324-8. https://doi.org/10.4315/0362-028X-61.3.324

Babic, I., & Watada, A.E. (1996). Microbial populations of fresh-cut spinach leaves affected by controlled atmospheres. *Postharvest Biol. Technol.*, 9, 187–193. https://doi.org/10.1016/S0925-5214(96)00047-6

Baiamonte, I., Raffo, A., Nardo, N., Moneta, E., Peparaio, M., D'Aloise, A., Kelderer, M., Casera, C., & Paoletti, F. (2016). Effect of the use of anti-hail nets on codling moth (*Cydia pomonella*) and organoleptic quality of apple (cv. Braeburn) grown in Alto Adige Region (northern Italy). *J. Sci. Food Agric.*, 96(6), 2025-2032. https://doi.org/10.1002/jsfa.7313

Baier, M., Foerster, J., Schnabel, U., Knorr, D., Ehlbeck, J., Herppich, W.B., & Schlüter, O. (2013). Direct non-thermal plasma treatment for the sanitation of fresh corn salad leaves: evaluation of physical and physiological effects and antimicrobial efficacy. *Postharvest Biol. Technol.*, 84, 81-87. https://doi.org/10.1016/j.postharvbio.2013.03.022

Baldwin, E.A., Scott, J.W., Shewmaker, C.K., & Schuch, W. (2000). Flavor trivia and tomato aroma: biochemistry and possible mechanisms for control of important aroma components. *HortSci.*, 35(6), 1013-1022. https://doi.org/10.21273/HORTSCI.35.6.1013

Baldwin, I.T., Kessler, A., & Halitschke, R. (2002). Volatile signaling in plant-plant-herbivore interactions: What is real? *J. Curr. Opin. Plant Biol.*, 5, 351–354. https://doi.org/10.1016/S1369-5266(02)00263-7

Baldwin, E.A., Plotto, A., & Goodner, K. (2007). Shelf-life versus flavour-life for fruits and vegetables: how to evaluate this complex trait. *Stewart Postharvest Rev.*, 3(1–2), 1–10. https://access.portico.org/stable?au=phx64r3rfk2

Bari, M.L., Nakauma, M., Todoriki, S., Juneja, V.K., Isshiki, K., & Kawamoto, S. (2005). Effectiveness of irradiation treatments in inactivating *Listeria monocytogenes* on fresh vegetables at refrigeration temperature. *J. Food Prot.*, 68, 318-323. https://doi.org/10.4315/0362-028X-68.2.318

Bartley, I.M., Stoker, P.G., Martin, A.D.E., Hatfield, S.G.S., & Knee, M. (1985). Synthesis of aroma com pounds by apples supplied with alcohols and methyl esters of fatty acids. *J. Sci. Food Agric.*, 36, 567-574. https://doi.org/10.1002/jsfa.2740360708

Barrett, D.M., Beaulieu, J.C., & Shewfelt, R.L. (2010). Color, flavor, texture and nutritional quality of fresh-cut fruits and vegetables: Desirable levels, instrumental and sensory measurement, and effects of processing. *Crit. Rev. Food Sci. Nutr.*, 50, 369–389. https://doi.org/10.1080/10408391003626322

Baselice, A., Colantuoni, F., Lass, D. A., Gianluca, N., & Stasi, A. (2015). EU consumers' perceptions of fresh-cut fruit and vegetables attributes: a choice experiment model. In *Selected Paper prepared for presentation at the Agricultural & Applied Economics Association's 2014* (pp. 27-29). Proc. AAEA An. Meeting, Minneapolis, USA. http://hdl.handle.net/11369/338363

Bauchot, A.D., Mottram, D.S., Dodson, A.T., & John, P. (1998). Effect of aminocyclopropane-1-carboxylic acid oxidase antisense gene on the formation of volatile esters in cantaloupe Charentais melon (cv. Vedrandais). *J. Agric. Food Chem.*, 46(11), 4787-4792. https://doi.org/10.1021/jf980692z

Beaudry, R.M. (2000). Response of horticultural commodities to low oxygen: limits to the expanded use of MAP. *Hort-Technol.,* *10*, 491–500. https://doi.org/10.21273/HORTTECH.10.3.491

Beaulieu, J.C. (2006). Effect of cutting and storage on acetate and nonacetate esters in convenient, ready-to-eat fresh-cut melons and apples. *HortSci.,* 41(1), 65-73. https://doi.org/10.21273/HORTSCI.41.1.65

Beaulieu, J. (2010). Factors affecting sensory quality of fresh-cut produce. In *Advances in Fresh-Cut Fruits and Vegetables* (pp. 115-143), Martin-Belloso, O., & Soliva Fortuny, R. (Eds.). CRC Press: Boca-Raton, FL, USA. https://doi.org/10.1201/b10263

Bett, K.L., Ingram, D.A., Grimm, C.C., Lloyd, S.W., Spanier, A.M., Miller, J.M., Gross, E.A., Baldwin, B.T., & Vinyard, B.T. (2001). Flavor of fresh-cut gala apples in barrier film packaging as affected by storage time. *J. Food Qual.*, 24(2), 141-156. https://doi.org/10.1111/j.1745-4557.2001.tb00597.x

Beuchat, L.R. (1996). Pathogenic microorganisms associated with fresh produce. *J. Food Prot.*, 59, 204–216. https://doi.org/10.4315/0362-028X-59.2.204

Beuchat, L.R., & Ryu, J.H. (1997). Produce handling and processing practices. *Emerg. Infect. Dis.*, 3, 459–465. https://doi.org/10.3201/eid0304.970407

Beuchat, L.F. (2000). Use of sanitizers in raw fruit and vegetable processing. In *Minimally processed fruits and vegetables* (pp. 63-78), Tapia, M.S., Lopez-Malo, A. & Alzamora, S.M. (Eds.). Aspen Publishers Inc: Gaithersburg, Maryland, USA.

Bhagwat, A.A. (2006). Microbiological safety of fresh-cut produce: where are we now? In *Microbiology of fresh produce* (pp. 121-165), Matthews, K., & Doyle, M. (Eds.). ASM Press: Washington, DC, USA. https://doi.org/10.1128/9781555817527.ch5

Biegańska-Marecik, R., & Czapski, J. (2007). The effect of selected compounds as inhibitors of enzymatic browning and softening of minimally processed apples. *Acta Sci. Pol. Technol. Aliment.,* *6*(3), 37-49. http://yadda.icm.edu.pl/yadda/element/bwmeta1.element.agro-article-195ea271-34cb-4bcd-a458-be75ea68ae67

Both, V., Brackmann, A., Thewes, F.R., de Freitas Ferreira, D., & Wagner, R. (2014). Effect of storage under extremely low oxygen on the volatile composition of 'Royal Gala'apples. *Food Chem.*, 156, 50-57. https://doi.org/10.1016/j.foodchem.2014.01.094

Bourne, M. (2002). *Food texture and viscosity: concept and measurement (4th Ed)*. Academic Press: New York, USA.

Brackett, R.E. (1997). Microbiological spoilage and pathogens in minimally processed refrigerated fruits and vegetables. In *Minimally processed refrigerated fruits & vegetables* (pp. 226–268), Wiley, R.C. (Ed.). Chapman and Hall: New York, USA. https://doi.org/10.1007/978-1-4615-2393-2_7

Brackmann, A., Streif, J., & Bangerth, F. (1993). Relationship between a reduced aroma production and lipid metabolism of apples after long-term controlled atmosphere storage. *J. Amer. Soc. Hort. Sci.*, 118, 243–247. https://doi.org/10.21273/JASHS.118.2.243

Brecht J.K. (2006). Controlled atmosphere, modified atmosphere and modified atmosphere packaging for vegetables. *Stewart Postharvest Rev.*, 2(5), 1–6. https://doi.org/10.2212/spr.2006.5.5

Buchanan, B.B., Gruissem, W., & Jones, R.L. (Eds.). (2015). *Biochemistry and molecular biology of plants (2nd ed)*. John Wiley & Sons: Somerset, NJ, USA.

Buchter-Weisbrodt, H. (1998). *Der Apfel*. Thieme Verlag, Stuttgart und Institut für Chemie und Biologie (ICB) in der ehemal. Bundesforschungsanstalt für Ernährung, Karlsruhe o.J.

Burdock, G.A. (2016). *Fenaroli's handbook of flavor ingredients (6th ed)*. CRC Press: Boca Raton, FL, USA. https://doi.org/10.1201/9781439847503

Burnett, S.L., & Beuchat, L.R. (2001). Human pathogens associated with raw produce and unpasteurized juices, and difficulties in decontamination. *J. Ind. Microbiol. Biotechnol.*, 27(2), 104-110. https://doi.org/10.1038/sj.jim.7000106

Busta, F.F., Suslow, T.V., Parish, M.E., Beuchat, L.R., Farber, J.N., Garett, E.H., & Harris, L.J. (2003). The use of indicators and surrogate microorganisms for the evaluation of pathogen in fresh and fresh cut produce. *Compr. Rev. Food Sci. Food Saf.*, 2, 179–185. https://doi.org/10.1111/j.1541-4337.2003.tb00035.x

Buta, J.G., Moline, H.E., Spaulding, D.W., & Wang, C.Y. (1999). Extending storage life of fresh-cut apples using natural products and their derivatives. *J. Agric. Food Chem.*, 47 (1), 1-6. https://doi.org/10.1021/jf980712x

Buttery, R.G. (1993). Quantitative and sensory aspects of flavor of tomato and other vegertables and fruits. In *Flavor Science: Sensible Principles and Techniques* (pp. 259–286). Acree, T.E., Teranishi, R. (Eds.). ACS: Washington, DC, USA.

Brückner B., & Wyllie. S.G. (2008). In *Fruit and Vegetable Flavour: Recent Advances and Future Prospects*. CRC Press, Woodhead: Cambridge, UK.

Caleb, O., Ilte, K., Fröhling, A., Geyer, M., Mahajan, P. (2006). Integrated modified atmosphere and humidity package design for minimally processed Broccoli (*Brassica oleracea* L. var. *italica*). *Postharvest Biol. Technol.*, 121, 87–100. https://doi.org/10.1016/j.postharvbio.2016.07.016

Caleb, O.J., Mahajan, P.V., Al-Said, F.A.J., & Opara, U.L. (2013). Modified atmosphere packaging technology of fresh and fresh-cut produce and the microbial consequences—a review. *Food Bioprocess Technol.*, 6(2), 303-329. https://doi.org/10.1007/s11947-012-0932-4

Caleb, O., Ilte, K., Fröhling, A., Geyer, M., & Mahajan, P. (2016). Integrated modified atmosphere and humidity package design for minimally processed broccoli (*Brassica oleracea* L. var. *italica*). *Postharvest Biol. Technol.*, 121, 87-100. https://doi.org/10.1016/B978-0-100596-5.21003-2

Capozzi, V., Fiocco, D., Amodio, M.L., Gallone, A., & Spano, G. (2009). Bacterial stressors in minimally processed food. *Int. J. Mol. Sci.*, 10, 3076–3105. https://doi.org/10.3390/ijms10073076

Castelló, M. L., Igual, M., Fito, P. J., & Chiralt, A. (2009). Influence of osmotic dehydration on texture, respiration and microbial stability of apple slices (var. Granny Smith). *J. Food Eng.*, 91(1), 1-9. https://doi.org/10.1016/j.jfoodeng.2008.07.025

Chand-Goyal, T., & Spotts, R.A. (1996). Enumeration of bacterial and yeast colonists of apple fruits and identification of epiphytic yeasts on pear fruits in the Pacific Northwest United States. *Microbiol. Res.*, 151, 427–432. https://doi.org/10.1016/S0944-5013(96)80013-9

Chardonnet, C.O., Charron, C.S., Sams, C.E., & Conway, W.S. (2003). Chemical changes in the cortical tissue and cell walls of calcium-infiltrated 'Golden Delicious' apples during storage. *Postharvest Biol. Technol.*, 28(1), 97-111. https://doi.org/10.1016/S0925-5214(02)00139-4

Charron, C.S., Cantliffe, D.J., & Heath, R.R. (1995). Volatile emissions from plants. *Hortic. Rev.*, 50, 43. https://doi.org/10.1002/9780470650585.ch2

Chen, C., Hu, W., He, Y., Jiang, A., & Zhang, R. (2016). Effect of citric acid combined with UV-C on the quality of fresh-cut apples. *Postharvest Biol. Technol.*, 111, 126-131. https://doi.org/10.1016/j.postharvbio.2015.08.005

Chiabrando, V., & Giacalone, G. (2012). Effect of antibrowning agents on color and related enzymes in fresh-cut apples during cold storage. *J. Food Process. Pres.*, 36(2), 133-140. https://doi.org/10.1111/j.1745-4549.2011.00561.x

Cliff, M.A., Toivonen, P.M., Forney, C.F., & Lu, C. (2010). Quality of fresh-cut apple slices stored in solid and micro-perforated film packages having contrasting O_2 headspace atmospheres. *Postharvest Biol. Technol.*, 58(3), 254-261. https://doi.org/10.1016/j.postharvbio.2010.07.015

Cocci, E., Rocculi, P., Romani, S., & Dalla Rosa, M. (2006). Changes in nutritional properties of minimally processed apples during storage. *Postharvest Biol. Technol.*, 39(3), 265-271. https://doi.org/10.1016/j.postharvbio.2005.12.001

Conesa, A., Verlinden, B.E., Artés-Hernández, F., Nicolaï, B., & Artés, F. (2007). Respiration rates of fresh-cut bell peppers under superamospheric and low oxygen with or without high carbon dioxide. *Postharvest Biol. and Technol.*, 45(1), 81-88. https://doi.org/10.1016/j.postharvbio.2007.01.011

Conte, A., Scrocco, C., Brescia, I., Mastromatteo, M., & DelNobile, M.A. (2011). Shelf life of fresh-cut Cime di rapa (*Brassica rapa* L.) as affected by packaging. *Food Sci. Technol.*, 44, 1218–1225. https://doi.org/10.1016/j.lwt.2010.11.006

Cortellino, G., Rizzolo, A., & Gobbi, S. (2013). Effect of conventional and alternative modified atmosphere packaging on the shelf-life of fresh-cut apples. In *XI International Controlled and Modified Atmosphere Research Conference,* 1071, 223-230. https://doi.org/10.17660/ActaHortic.2015.1071.25

Cortellino, G., Gobbi, S., Bianchi, G., & Rizzolo, A. (2015). Modified atmosphere packaging for shelf life extension of fresh-cut apples. *Trends Food Sci. Tech.*, 46(2), 320–330. https://doi.org/10.1016/j.tifs.2015.06.002

Coupe, S.A., Sinclair, B.K., Watson, L.M., Heyes, J. A., & Eason, J.R. (2003). Identification of dehydration-responsive cysteine proteases during post-harvest senescence of broccoli florets. *J. Exp. Bot.*, 54(384), 1045-1056. https://doi.org/10.1093/jxb/erg105

Cunningham, D.G, Acree, T.E., Barnard, J., Butts, R.M., & Breall, P.A. (1986). Charm analysis of apple volatiles. *Food Chem.,* 137-147. https://doi.org/10.1016/0308-8146(86)90107-X

De Pooter, H.L., Dirinck, P.J., Willaert, G.A., & Schamp, N.M. (1981). Metabolism of propionic acid by Golden Delicious apples. *Phytochemistry*, 20 (9), 2135-2138. https://doi.org/10.1016/0031-9422(81)80100-8

De Pooter, H.L., Montens, J.P., Willaert, G.A., Dirinck, P.J., & Schamp, N.M. (1983). Treatment of Golden Delicious apples with aldehydes and carboxylic acids: effect on the headspace composition. *J. Agric. Food Chem.,* 31(4), 813-818. https://doi.org/10.1021/jf00118a034

Defilippi, B.G., Dandekar, A.M., & Kader, A.A. (2005). Relationship of ethylene biosynthesis to volatile production, related enzymes, and precursor availability in apple peel and flesh tissues. *J. Agric. Food Chem.,* 53(8), 3133-3141. https://doi.org/10.1021/jf047892x

Defilippi, B.G., Manriquez, D., Luengwilai, K., & González-Agüero, M. (2009). Aroma volatiles: biosynthesis and mechanisms of modulation during fruit ripening. *Adv. Bot. Res.*, 50, 1-37. https://doi.org/10.1016/S0065-2296(08)00801-X

DeRovira, D., & Mermelstein, N. (1996). The dynamic flavor profile method. *Food Technol. (Chicago)*, 50(2), 55-60.

Devlieghere, F., & Debevere, J. (2000). Influence of dissolved carbon dioxide on the growth of spoilage bacteria. *LWT-Food Sci. Technol.*, 33(8), 531-537. https://doi.org/10.1006/fstl.2000.0705

DGHM (2011) Microbiological guiding and warning limits for food stuffs. German Association for Hygiene and Microbiology: Hannover, Germany. https://www.dghm-richt-warnwerte.de/de

Dickinson, J.R., Harrison, S.J., Dickinson, J.A., & Hewlins, M.J. (2000). An investigation of the metabolism of isoleucine to active amyl alcohol in Saccharomyces cerevisiae. *J. Biol. Chem.,* 275(15), 10937-10942. https://doi.org/10.1074/jbc.275.15.10937

Diebold, R., Schuster, J., Däschner, K., & Binder, S. (2002). The branched-chain amino acid transaminase gene family in Arabidopsis encodes plastid and mitochondrial proteins. *Plant Physiol.,* 129(2), 540-550. https://doi.org/10.1104/pp.001602

Dimick, P.S., & Hoskin, J.C. (1983). Review of apple flavor—state of the art. *Crit. Rev. Food Sci. Nutr.*, 18, 387-409. https://doi.org/10.1080/10408398309527367

Dixon, J., & Hewett, E.W. (2000). Factors affecting apple aroma/flavour volatile concentration: a review. *N. Z. J. Crop Hortic. Sci.,* 28(3), 155-173. https://doi.org/10.1080/01140671.2000.9514136

Dürr, P., & Schobinger, U. (1981). The concentration of some volatiles to the sensory quality of apple and orange juice odour. In *Flavour'81 3rd Weurman Symposium Proceedings of the International Conference* (pp. 179-193). Munich April 28-30. Schreier, P. (Ed.). Walter de Gruyter GmbH & Co KG: Berlin, New York.

Drawert, F., Heimann, W., Emberger, R., & Tressl, R. (1969). Über die Biogenese von Aromastoffen bei Pflanzen und Früchten IV. Mitteilung Bildung der Aromastoffe des Apfels im Verlauf des Wachstums und bei der Lagerung. *Zeitschrift für Lebensmittel-Untersuchung und Forschung*, 140(2), 65-88. https://doi.org/10.1007/BF01387242

Eberhardt, M.V., Lee, C.Y., & Liu, R.H. (2000). Antioxidant activity of fresh apples. *Nature*, 405(6789), 903-904. https://doi.org/10.1038/35016151

Echeverrıa, G., Fuentes, T., Graell, J., Lara, I., & López, M.L. (2004). Aroma volatile compounds of 'Fuji'apples in relation to harvest date and cold storage technology: A comparison of two seasons. *Postharvest Biol. Technol.*, 32(1), 29-44. https://doi.org/10.1016/j.postharvbio.2003.09.017

Eisenreich, W., Bacher, A., Arigoni, D., & Rohdich, F. (2004). Biosynthesis of isoprenoids via the non-mevalonate pathway. *Cell. Mol. Life Sci.*, 61(12), 1401-1426. https://doi.org/10.1007/s00018-004-3381-z

El Hadi, M.A.M., Zhang, F.J., Wu, F.F., Zhou, C.H., & Tao, J. (2013). Advances in fruit aroma volatile research. *Molecules*, 18(7), 8200-8229. https://doi.org/10.3390/molecules18078200

Escartin, E.F., Ayala, A.C., & Lozano, J.S. (1989). Survival and growth of *Salmonella* and *Shigella* on sliced fresh fruit. *J. Food Prot.*, 52(7), 471-472. https://doi.org/10.4315/0362-028X-52.7.471

Espino-Díaz, M., Sepúlveda, D.R., González-Aguilar, G., & Olivas, G.I. (2016). Biochemistry of apple aroma: A review. *Food Technol. Biotech.,* 54(4), 375-394. https://doi.org/10.17113/ftb.54.04.16.4248

European Commission. (2005). Commission Regulation (EC) No 2073/2005 of 15 November 2005 on microbiological criteria for foodstuffs. *Official Journal of the European Union*, 338, 1-26. http://data.europa.eu/eli/reg/2005/2073/oj

Erkan M., & Wang, C.Y. (2006). Modified and controlled atmosphere storage of subtropical crops. *Stewart Postharvest Rev.*, 2(5-4), 1–8. https://doi.org/10.2212/spr.2006.5.4

Fallik, E., Archbold, D.D., Hamilton-Kemp, T.R., Loughrin, J.H., & Collins, R.W. (1997). Heat treat ment temporarily inhibits aroma volatile com pound emission from Golden Delicious apples. *J. Agric. Food Chem.,* 45, 4038-4041. https://doi.org/10.1021/jf970358n

Fallik, E. (2004). Prestorage hot water treatments (immersion, rinsing and brushing). *Postharvest Biol. Technol.*, 32, 125–134. https://doi.org/10.1016/j.postharvbio.2003.10.005

Fan, L., Song, J., Forney, C.F., & Jordan, M.A. (2005). Ethanol production and chlorophyll fluorescence predict breakdown of heat-stressed apple fruit during cold storage. *J. Am. Soc. Hortic. Sci.*, 130(2), 237-243. https://doi.org/10.21273/JASHS.130.2.237

Fan, L., Song, J., Forney, C.F., & Jordan, M.A. (2011). Fruit maturity affects the response of apples to heat stress. *Postharvest Biol. Technol.*, 62, 35–42. https://doi.org/10.1016/j.postharvbio.2011.04.007

Farber, J.M. (1991). Microbiological aspects of modified-atmosphere packaging technology – A review. *J. Food Prot.,* 54(1), 58–70. https://doi.org/10.4315/0362-028X-54.1.58

Farber, J.N., Harris, L.J., Parish, M.E., Beuchat, L.R., Suslow, T.V., Gorney, J.R., Garrett, E.H., & Busta, F.F. (2003). Microbiological safety of controlled and modified atmosphere packaging of fresh and fresh-cut produce. *Compr. Rev. Food Sci. Food Saf.*, 2, 142-160. https://doi.org/10.1111/j.1541-4337.2003.tb00032.x

Fellman, J.K., Mattheis, J.P., Matthinson, D.S., & Bostick, B.C. (1991). Assay of acetyl-CoA alcohol transferase in 'Delicious' apples. *HortSci.*, 27, 773-776.

Fellman, J.K., Mattinson, D.S., Bostick, B.C., Mattheis, J.P., & Patterson, M.E. (1993). Ester biosynthesis in 'Rome' apples subjected to low-oxygen atmospheres. *Postharvest Biol. Technol.*, 3(3), 201-214. https://doi.org/10.1016/0925-5214(93)90056-9

Fellman, J.K., Miller, T.W., Mattinson, D.S., & Mattheis, J.P. (2000). Factors that influence biosynthesis of volatile flavor compounds in apple fruits. *HortSci.,* 35(6), 1026-1033. https://doi.org/10.21273/HORTSCI.32.3.554C

Ferreira, L., Perestrelo, R., Caldeira, M., & Câmara, J.S. (2009). Characterization of volatile substances in apples from Rosaceae family by headspace solid-phase microextraction followed by GC-qMS. *J. Sep. Sci.*, 32(11), 1875-1888. https://doi.org/10.1002/jssc.200900024

Finnegan, E., Mahajan, P.V., O'Connell, M., Francis, G.A., & O'Beirne, D. (2013). Modelling respiration in fresh-cut pineapple and prediction of gas permeability needs for optimal modified atmosphere packaging. *Postharvest Biol. Technol.*, 79, 47-53. https://doi.org/10.1016/j.postharvbio.2012.12.015

Fischer, R.L., & Bennett, A.B. (1991). Role of cell wall hydrolases in fruit ripening. *Annu. Rev. Plant Biol.*, 42(1), 675-703. https://doi.org/10.1146/annurev.pp.42.060191.003331

Flath, R.A., Black, D.R., GuActagni, D.G., McFadden, W.H., & Schultz, T.H. (1967). Identification and organoleptic evaluation of compounds in Delicious apple essence. *J. Agric. Food Chem.*, 15, 29-35. https://doi.org/10.1021/jf60149a032

Fonseca, S.C., Oliveira, F.A., & Brecht, J.K. (2002). Modelling respiration rate of fresh fruits and vegetables for modified atmosphere packages: a review. *J. Food Eng.*, *52*(2), 99-119. https://doi.org/10.1016/S0260-8774(01)00106-6

Forney, C.F., Kalt, W., & Jordan, M.A. (2000). The composition of strawberry aroma is influenced by cultivar, maturity, and storage. *HortSci.,* 35(6), 1022-1026. https://doi.org/10.21273/HORTSCI.32.3.554D

Forney, C.F. (2008). Flavour loss during postharvest handling and marketing of fresh-cut produce. *Stewart Postharvest Rev.*, 4(3-5), 1-10. https://doi.org/10.2212/spr.2008.3.5

Frenkel, C., & Patterson, M.E. (1973). Effect of carbon dioxide on the activity of succinic dehydrogenase in 'Bartlett' pears during cold storage. *HortSci.*, 8, 395–396.

Gacche, R.N., Shete, A.M., Dhole, N.A., & Ghole, V.S. (2006). Reversible inhibition of polyphenol oxidase from apple using L-cysteine. *Indian J. Chem. Technol.*, 13(5), 459-463. http://nopr.niscair.res.in/handle/123456789/7093

Galliard, T. (1968). Aspects of lipid metabolism in higher plants—II. The identification and quantitative analysis of lipids from the pulp of pre- and post climacteric apples. *Phytochemistry*, 7, 1915-1922. https://doi.org/10.1016/S0031-9422(00)90751-9

Gang, D.R., Wang, J., Dudareva, N., Nam, K.H., Simon, J.E., Lewinsohn, E., & Pichersky, E. (2001). An investigation of the storage and biosynthesis of phenylpropenes in sweet basil. *Plant Physiol.*, 125(2), 539-555. https://doi.org/10.1104/pp.125.2.539

Garcia, E., & Barrett, D.M. (2005). Fresh-cut fruits. In *Processing fruits* (*2nd Ed.*, pp. 53-72), Barrett, D.M., Somogyi, L., & Ramaswamy, H.S. (Eds.). CRC Press: Boca Raton, Fla, USA. https://doi.org/10.1201/9781420040074

Garcia, E., & Barrett, D.M. (2002). Preservative treatments for fresh-cut fruits and vegetables. In *Fresh-Cut Fruits and Vegetables: Science, Technology, and Market* (pp. 267-304). Lamikanra, O. (Ed.). CRC Press LLC: Boca Raton, Florida, USA. https://doi.org/10.1201/9781420031874

Gardner, H.W. (1995). Biological roles and biochemistry of the lipoxygenase pathway. *HortSci.*, 30(2), 197-205.

Goepfert, S., & Poirier, Y. (2007). β-Oxidation in fatty acid degradation and beyond. *Curr. Opin. Plant Biol.*, 10(3), 245-251. https://doi.org/10.1016/j.pbi.2007.04.007

Gil, M.I., Gorny, J.R., & Kader, A.A. (1998). Responses of 'Fuji' apple slices to ascorbic acid treatments and low-oxygen atmospheres. *HortSci.*, 33(2), 305-309.

Golden, D.A., Rhodehamel, E.J., & Kautter, D.A. (1993). Growth of *Salmonella* spp. in cantaloupe, watermelon, and honeydew melons. *J. Food Prot.*, 56, 194–196. https://doi.org/10.4315/0362-028X-56.3.194

Goodenough, P.W. (1983). Increase in esterase as a func tion of apple fruit ripening. *Acta Hortic.*, 138, 83-92. https://doi.org/10.17660/ActaHortic.1983.138.9

Gorny, J.R., Hess-Pierce, B., & Kader, A.A. (1998). Effects of Fruit Ripeness and Storage. *HortSci.*, 33(1), 110-113.

Gorny, J.R., Cifuentes, R.A., Hess-Pierce, B., & Kader, A.A. (2000). Quality changes in fresh-cut pear slices as affected by cultivar, ripeness stage, fruit size, and storage regime. *Int. J. Food Sci. Technol.*, 65, 541–544. https://doi.org/10.1016/S0925-5214(01)00139-9

Gorny, J.R., (2001). A summary of CA and MA requirements and recommendations for fresh-cut (minimally processed) fruits and vegetables. *Acta Hortic.*, 600, 609-614. https://doi.org/10.17660/ActaHortic.2003.600.92

Gorny J.R., Hess-Pierce, B., Cifuentes, R.A., & Kader, A.A. (2002). Quality changes in fresh cut pear slices as affected by controlled atmospheres and chemical preservatives. *Postharvest Biol. Technol.,* 24(3), 271–278. https://doi.org/10.1016/S0925-5214(01)00139-9

Graça, A., Santo, D., Esteves, E., Nunes, C., Abadias, M., & Quintas, C. (2015). Evaluation of microbial quality and yeast diversity in fresh-cut apple. *Food Microbiol.*, 51, 179-185. https://doi.org/10.1016/j.fm.2015.06.003

Graham, I.A., & Eastmond, P.J. (2002). Pathways of straight and branched chain fatty acid catabolism in higher plants. *Prog. Lipid Res.,* 41(2), 156-181. https://doi.org/10.1016/S0163-7827(01)00022-4

Guadagni, D.G., Bomben, J.L., & Hudson, J.S. (1971). Factors influencing the development of aroma in apple peels. *J. Sci. Food Agric.*, 22, 110-114. https://doi.org/10.1002/jsfa.2740220303

Gunes, G., Watkins, C.B., & Hotchkiss, J.H. (2001). Physiological responses of fresh-cut apple slices under high CO_2 and low O_2 partial pressures. *Postharvest Biol. Technol.,* 22(3), 197-204. https://doi.org/10.1016/S0925-5214(01)00083-7

Hansen, K., & Poll, L. (1993). Conversion of L-isoleucine into 2-methylbut-2-enyl esters in apples. *Food Sci. Technol.*, 26, 178–180. https://doi.org/10.1006/fstl.1993.1036

Heard, G.M. (1999). Microbial safety of ready-to-eat salads and minimally processed vegetables and fruits. *Food Aust.*, 51, 414–420.

Heard, G.M. (2002). Microbiology of fresh-cut produce. In *Fresh-cut fruits and vegetables: Science, technology, and market* (pp. 187-249). Lamikanra, O. (Ed.). CRC Press LLC: Boca Raton, Florida, USA. https://doi.org/10.1201/9781420031874

Hecke, K., Herbinger, K., Veberič, R., Trobec, M., Toplak, H., Štampar, F., Keppel, H., & Grill, D. (2006). Sugar-, acid- and phenol contents in apple cultivars from organic and integrated fruit cultivation. *Eur. J. Clin. Nutr.*, 60(9), 1136-1140. https://doi.org/10.1038/sj.ejcn.1602430

Heinrich, V., Zunabovic, M., Nehm, L., Bergmair, J., & Kneifel, W. (2016). Influence of argon modified atmosphere packaging on the growth potential of strains of *Listeria monocytogenes* and *Escherichia coli*. *Food Control*, 59, 513-523. https://doi.org/10.1016/j.foodcont.2015.06.010

Hengari, S., Theron, K.I., & Steyn, W.J. (2016). Differential dependence of apple (*Malus domestica* Borkh.) cultivars on the xanthophyll cycle for photoprotection. S. Afr. J. Plant Soil, 33(1), 69-76. https://doi.org/10.1080/02571862.2015.1056849

Herppich, W.B., Flach, B.M.T., Von Willert, D.J., & Herppich, M. (1997). Field investigations in *Welwitschia mirabilis* during a severe drought.: II. Influence of leaf age, leaf temperature and irradiance on photosynthesis and photoinhibition. *Flora*, 192(2), 165-174. https://doi.org/10.1016/S0367-2530(17)30773-9

Herppich, W.B., Herold, B., Geyer, M., & Gomez, F. (2004). Effects of temperature and water relations on carrots and radish tuber texture. *J. Appl. Bot.*, 78, 11–17. https://lup.lub.lu.se/record/139175

Herppich, W.B., Huyskens-Keil, S., & Kadau, R. (2005). Effects of short-term low-temperature storage on mechanical and chemical properties of white asparagus cell walls. *J. Appl. Bot. Food Qual.*, 79, 63. https://www.researchgate.net/publication/280727195_Effects_of_short-term_low-temperature_storage_on_mechanical_and_chemical_properties_of_white_Asparagus_cell_walls

Herppich, W.B., Maggioni, M., Huyskens-Keil, S., Kabelitz, T., & Hassenberg, K. (2020). Optimization of Short-Term Hot-Water Treatment of Apples for Fruit Salad Production by Non-Invasive Chlorophyll-Fluorescence Imaging. *Foods*, 9, 820. https://doi.org/10.3390/foods9060820

Hodges D.M., & Toivonen, P.M.A. (2008). Quality of fresh-cut fruits and vegetables as affected by exposure to abiotic stress. *Postharvest Biol. Technol*, 48(2), 155–162. https://doi.org/10.1016/j.postharvbio.2007.10.016

Iqbal, T., Rodrigues, F.A.S., Mahajan, P.V., Kerry, J.P., Gil, L., Manso, M.C., & Cunha, L.M. (2008). Effect of minimal processing conditions on respiration rate of carrots. *J. Food Sci.*, 73(8), E396-E402. https://doi.org/10.1111/j.1750-3841.2008.00923.x

James, J.B., Ngarmsak, T., & Rolle, R.S. (2010). Processing of fresh-cut tropical fruits and vegetables: A technical guide. *RAP Publication (FAO) eng no. 2010/16.*

Jay, J.M. (1986). The microbial spoilage of foods. In *Perspectives in Biotechnology and Applied Microbiology* (pp. 325-342), Alani, D.I., & Moo-Young, M. (Eds.). Springer: Dordrecht, Netherlands. https://doi.org/10.1007/978-94-009-4321-6

Jiang, Y., Pen, L., & Li, J. (2004). Use of citric acid for shelf life and quality maintenance of fresh-cut Chinese water chestnut. *J. Food Eng.*, 63(3), 325-328. https://doi.org/10.1016/j.jfoodeng.2003.08.004

Kabelitz, T., & Hassenberg, K. (2018). Control of apple surface microflora for fresh-cut produce by post-harvest hot-water treatment. *LWT-Food Sci. Technol.*, 98, 492–499. https://doi.org/10.1016/j.lwt.2018.08.062

Kabelitz, T., Schmidt, B., Herppich, W.B., & Hassenberg, K. (2019). Effects of hot water dipping on apple heat transfer and post-harvest fruit quality. *LWT-Food Sci. Technol.*, 108, 416–420. https://doi.org/10.1016/j.lwt.2019.03.067

Kader, A.A. (1987). Respiration and gas exchange of vegetables. In *Postharvest physiology of vegetables* (pp. 25-43), Weichmann, J. (Ed.) Marcel Dekker: New York, USA.

Kader, A.A. (2002). Postharvest biology and technology: An overview. In *Postharvest technology of horticultural crops Vol. 3311* (pp. 39–48). Kader, A.A. (Ed.). University of California, Division of Agriculture and Natural Resources: Davis, CA, USA.

Kader, A.A., & Saltveit, M.E. (2003). Respiration and gas exchange. In *Postharvest physiology and pathology of vegetables (2nd Ed.,* pp. 7-29*)*. Bartz, J.A., Brecht, J.K. (Eds.). CRC Press: Boca Raton, FL, USA. https://doi.org/10.1201/9780203910092

Kader, A.A. (2008). Flavor quality of fruits and vegetables. *J. Sci. Food Agric.*, 88(11), 1863-1868. https://doi.org/10.1002/jsfa.3293

Kader A.A. (2010). Future of modified atmosphere research. *Acta Hortic.*, 857, 213-217. https://doi.org/10.17660/ActaHortic.2010.857.24

Kadereit, J.W., Körner, Ch., Kost, B., & Sonnewald, U. (2014). *Strasburger – Lehrbuch der Pflanzenwissenschaften (37th Ed)*. Springer-Verlag: Berlin, Heidelberg, Germany. https://doi.org/10.1007/978-3-662-61943-8

Karaibrahimoglu, Y., Fan, X., Sapers, G.M., & Sokorai, K. (2004). Effect of pH on the survival of *Listeria innocua* in calcium ascorbate solutions and on quality of fresh-cut apples. *J. Food Prot.*, 67, 751–757. https://doi.org/10.4315/0362-028X-67.4.751

Ke, D., Yahia, E., Mateos, M., & Kader, A.A. (1994). Ethanolic fermentation of Bartlett' pears as influenced by ripening stage and atmospheric composition. *J. Am. Soc. Hortic. Sci.,* 119(5), 976-982. https://doi.org/10.21273/JASHS.119.5.976

Kim, D.M., Smith, N.L., & Lee, C.Y. (1993). Apple cultivar variations in response to heat treatment and minimal processing. *J. Food Sci.*, 58, 1111–1114. https://doi.org/10.1111/j.1365-2621.1993.tb06126.x

Klein, J.D. (1989). Ethylene biosynthesis in heat-treated apples. In *Biochemical and physiological aspects of ethylene production in lower and higher plants* (pp. 181-189), Clijsters, H., de Proft, M., Marcelle, R., van Poucke, M., (Eds.). Springer: Dordrecht, Netherlands. https://doi.org/10.1007/978-94-009-1271-7_21

Knee, M., & Hatfield, S.G. (1981). The metabolism of alcohols by apple fruit tissue. *J. Sci. Food Agric.*, 32 (6), 593-600. https://doi.org/10.1002/jsfa.2740320611

Kollmannsberger, H., & Berger, R.G. (1992). Precursor atmosphere storage induced flavour changes in apples cv. Red Delicious. *Chem. Mikrobiol. Technol. Lebensm.*, 14, 81-86.

Koukounaras, A., Diamantidis, G., & Sfakiotakis, E. (2008). The effect of heat treatment on quality retention of fresh-cut peach. *Postharvest Biol. Technol.*, 48(1), 30-36. https://doi.org/10.1016/j.postharvbio.2007.09.011

Kurenda, A., Zdunek, A., Schlüter, O., & Herppich, W.B. (2014). VIS/NIR spectroscopy, chlorophyll fluorescence, biospeckle and backscattering to evaluate changes in apples subjected to hydrostatic pressures. *Postharvest Biol. Technol.*, 96, 88–98. https://doi.org/10.1016/j.postharvbio.2014.05.009

Lakakul, R., Beaudry, R.M., & Hernandez, R.J. (1999). Modeling respiration of apple slices in modified-atmosphere packages. *J. Food Sci.*, 64(1), 105-110. https://doi.org/10.1111/j.1365-2621.1999.tb09870.x

Lamikanra, O., & Watson, M.A. (2000). Cantaloupe melon peroxidase: characterization and effects of food additives on activity. *Nahrung/Food*, 44(3), 168–172. https://doi.org/10.1002/1521-3803(20000501)44:3<168::AID-FOOD168>3.0.CO;2-H

Lamikanra, O., Chen, J.C., Banks, D., & Hunter, P.A. (2000). Biochemical and microbial changes during the storage of minimally processed cantaloupe. *J. Agric. Food Chem.*, 48(12), 5955-5961. https://doi.org/10.1021/jf0000732

Lanciotti, R., Corbo, M.R., Gardini, F., Sinigaglia, M., & Guerzoni, M.E. (1999). Effect of hexanal on the shelf life of fresh apple slices. *J. Agric. Food Chem.*, 47(11), 4769–4776. https://doi.org/10.1021/jf990611e

Lara, I., Graell, J., López, M.L., & Echeverría, G. (2006). Multivariate analysis of modifications in biosynthesis of volatile compounds after CA storage of 'Fuji' apples. *Postharvest Biol. Technol.*, 39(1), 19-28. https://doi.org/10.1016/j.postharvbio.2005.09.001

Larsen, M., & Watkins, C.B. (1995). Firmness and concentrations of acetaldehyde, ethyl acetate and ethanol in strawberries stored in controlled and modified atmospheres. *Postharvest Biol. Technol.*, 5(1-2), 39-50. https://doi.org/10.1016/0925-5214(94)00012-H

Laurila, E., Kervinen, R., & Ahvenainen, R. (1998). The inhibition of enzymatic browning in minimally processed vegetables and fruits. *Postharvest news and information*, 9(4), 53-66. https://ucanr.edu/datastoreFiles/608-361.pdf

Lawlor, K.A., Schuman, J.D., Simpson, P.G., & Taormina, P.J. (2009). Microbiological spoilage of beverages. In *Compendium of the microbiological spoilage of foods and beverages* (pp. 245-284), Sperber, W.H., Doyle, M.P. (Eds.). Springer: New York, USA. https://doi.org/10.1007/978-1-4419-0826-1_9

Leverentz, B., Conway, W.S., Alavidze, Z., Janisiewicz, W.J., Fuchs, Y., Camp, M.J., Chighladze, E., & Sulakvelidze, A. (2001). Examination of bacteriophage as a biocontrol method for Salmonella on fresh-cut fruit-a model study. *J. Food Prot.*, 64, 1116–1121. https://doi.org/10.4315/0362-028X-64.8.1116

Li, W. L., Li, X.H., Sun, Y.J., Tang, Y., Jiang, Y.Q., & Zhang, M. (2011). Effect of packaging conditions on physiology quality and shelf-life of fresh-cut kiwifruit. In *Advanced Materials Research* (Vol. 233, pp. 1985-1988). Trans Tech Publications Ltd: Switzerland. https://doi.org/10.4028/www.scientific.net/AMR.233-235.1985

Liu, M., Nauta, A., Francke, C., & Siezen, R.J. (2008). Comparative genomics of enzymes in flavor-forming pathways from amino acids in lactic acid bacteria. *Appl. Environ. Microbiol.*, 74(15), 4590-4600. https://doi.org/10.1128/AEM.00150-08

López-Rubira, V., Conesa, A., Allende, A., & Artés, F. (2005). Shelf life and overall quality of minimally processed pomegranate arils modified atmosphere packaged and treated with UV-C. *Postharvest Biol. Technol.*, 37(2), 174-185. https://doi.org/10.1016/j.postharvbio.2005.04.003

Lozano-De-Gonzalez, P.G., Barrett, D.M., Wrolstad, R.E., & Durst, R.W. (1993). Enzymatic browning inhibited in fresh and dried apple rings by pineapple juice. *J. Food Sci.*, 58(2), 399-404. https://doi.org/10.1111/j.1365-2621.1993.tb04284.x

Luo, Y., & Barbosa-Canovas, G.V. (1996). Preservation of apple slices using ascorbic acid and 4-hexylresorcinol. *Food Sci. Technol. Int.*, 2(5), 315-321. https://doi.org/10.1177/108201329600200505

Luo, Y., & Barbosa-Cánovas, G.V. (1997). Enzymatic browning and its inhibition in new apple cultivars slices using 4-hexylresorcinol in combination with ascorbic acid. *Food Sci. Technol. Int.*, 3(3), 195-201. https://doi.org/10.1177/108201329700300307

Lurie, S., Fallik, E., & Klein, J.D. (1996). The effect of heat treatment on apple epicuticular wax and calcium uptake. *Postharvest Biol. Technol.*, 8, 271–277. https://doi.org/10.1016/0925-5214(96)00007-5

Lurie, S. (1998). Postharvest heat treatments. *Postharvest Biol. Technol.*, 14, 257–269. https://doi.org/10.1016/S0925-5214(98)00045-3

Mahajan, P.V., Lucas, A., & Edelenbos, M. (2014). Impact of mixtures of different fresh-cut fruits on respiration and ethylene production rates. *J. Food Sc.*, 79(7), 1366-1371. https://doi.org/10.1111/1750-3841.12512

Marilley, L., & Casey, M.G. (2004). Flavours of cheese products: metabolic pathways, analytical tools and identification of producing strains. *Int. J. Food Microbiol.*, 90(2), 139-159. https://doi.org/10.1016/S0168-1605(03)00304-0

Martinez, M.V., & Whitaker, J.R. (1995). The biochemistry and control of enzymatic browning. *Trends Food Sci. Technol.*, 6(6), 195-200. https://doi.org/10.1016/S0924-2244(00)89054-8

Martin-Belloso, O., Soliva-Fortuny, R., & Oms-Oliu, G. (2006). Freshcut fruits. In, Handbook of fruits and fruit processing (pp. 129–144), Hui, Y.H. (Ed.). Wiley-Blackwell: Lowa, CT.

Maskan, M. (2001). Kinetics of colour change of kiwifruits during hot air and microwave drying. *J. Food Eng.*, 48(2), 169-175. https://doi.org/10.1016/S0260-8774(00)00154-0

Mathews, C.K., & van Holde, KE. (1996). *Biochemistry 2nd edition* (pp. 1159). The Benjamin/Cummings Publishing Company, Inc.

Matich, A., & Rowan, D. (2007). Pathway analysis of branched-chain ester biosynthesis in apple using deuterium labeling and enantioselective gas chromatography− mass spectrometry. *J. Agric. Food Chem.,* 55(7), 2727-2735. https://doi.org/10.1021/jf063018n

Mattheis, J.P., Buchanan, D.A., & Fellman, J.K. (1991). Change in apple fruit volatiles after storage in atmospheres inducing anaerobic metabolism. *J. Agric. Food Chem.*, 39(9), 1602-1605. https://doi.org/10.1021/jf00009a012

Mattheis, J.P., Fan, X., & Argenta, L.C. (2005). Interactive responses of Gala apple fruit volatile production to controlled atmosphere storage and chemical inhibition of ethylene action. *J. Agric. Food Chem.,* 53(11), 4510-4516. https://doi.org/10.1021/jf050121o

Matyssek, R., & Herppich, W.B. (2018) Chlorophyllfluoreszenzanalyse. In *Experimentelle Pflanzenökologie* (pp. 1-56). Springer Reference Naturwissenschaften. Springer Spektrum: Berlin, Heidelberg, Germany. https://doi.org/10.1007/978-3-662-53493-9_13-1

Maxin, P., Weber, R.W., Pedersen, H.L., & Williams, M. (2012a). Hot-water dipping of apples to control *Penicillium expansum*, *Neonectria galligena* and *Botrytis cinerea*: Effects of temperature on spore germination and fruit rots. *Eur. J. Hortic. Sci.*, 77, 1–9.

Maxin, P., Weber, R.W., Pedersen, H.L., & Williams, M. (2012b). Control of a wide range of storage rots in naturally infected apples by hot-water dipping and rinsing. *Postharvest Biol. Technol.*, 70, 25–31. https://doi.org/10.1016/j.postharvbio.2012.04.001

Maxin, P., Williams, M., & Weber, R.W. (2014). Control of fungal storage rots of apples by hot-water treatments: A Northern European perspective. *Erwerbs-Obstbau*, 56, 25–34. http://dx.doi.org/10.1007/s10341-014-0200-z

Montero-Calderón, M., Rojas-Graü, M.A., & Martín-Belloso, O. (2008). Effect of packaging conditions on quality and shelf-life of fresh-cut pineapple (*Ananas comosus*). *Postharvest Biol. Technol.*, 50, 182–189. https://doi.org/10.1016/j.postharvbio.2008.03.014

Mpelasoka, B.S., & Behboudian, M.H. (2002). Production of aroma volatiles in response to deficit irrigation and to crop load in relation to fruit maturity for 'Braeburn' apple. *Postharvest Biol. Technol.*, 24(1), 111-116. https://doi.org/10.1016/S0925-5214(01)00110-7

Nguyen-the, C., & Carlin, F. (1994). The microbiology of minimally processed fresh fruit and vegetables. *Crit. Rev. Food Sci. Nutr.*, 34(4), 371–401. https://doi.org/10.1080/10408399409527668

Nicolas, J.J., Richard-Forget, F.C., Goupy, P.M., Amiot, M.J., & Aubert, S.Y. (1994). Enzymatic browning reactions in apple and apple products. *Crit. Rev. Food Sci. Nutr.*, *34*(2), 109-157. https://doi.org/10.1080/10408399409527653

Nicoli, M.C., Anese, M., & Severini, C. (1994). Combined effects in preventing enzymatic browning reactions in minimally processed fruit. *J. Food Qual.*, 17(3), 221-229. https://doi.org/10.1111/j.1745-4557.1994.tb00145.x

Nijssen, L.M., van Ingen-Visscher, C.A., & Donders, J.J.H. (2011). VCF Volatile Compounds in Food: database (Version 13.1.). Zeist (The Netherlands.

Nishikawa, F., Iwama, T., Kato, M., Hyodo, H., Ikoma, Y., & Yano, M. (2005). Effect of sugars on ethylene synthesis and responsiveness in harvested broccoli florets. *Postharvest Biol. Technol.*, 36(2), 157-165. https://doi.org/10.1016/j.postharvbio.2004.12.001

NIST Chemistry WebBook (2018). In *NIST Standard reference database number 69*. Linstrom, P.J., & Mallard, W.G. (Eds.). National Institute of Standards and Technology: Gaithersburg, MD, USA. https://doi.org/10.18434/T4D303 (last accessed on 25 August 2019).

Obando-Ulloa, J.M., Nicolai, B., Lammertyn, J., Bueso, M.C., Monforte, A.J., & Fernández-Trujillo, J.P. (2009). Aroma volatiles associated with the senescence of climacteric or non-climacteric melon fruit. *Postharvest Biol. Technol.*, 52(2), 146-155. https://doi.org/10.1016/j.postharvbio.2008.11.007

Paillard, N.M. (1979). Biosynthese des produits volatils de la pomme: formation des alcools et des esters a partir des acides gras. *Phytochemistry*, 18(7), 1165-1171. https://doi.org/10.1016/0031-9422(79)80127-2

Paillard, N.M.M. (1990). The flavour of apples, pears and quinces. In *Food flavours, Part C. The flavour of fruits* (pp. 1-41), Morton, I.D., MacLeod, A.J. (Eds.). Elsevier Science Publishing Company Inc.: Amsterdam, Netherlands.

Pateraki, I., & Kanellis, A.K. (2010). Stress and developmental responses of terpenoid biosynthetic genes in *Cistus creticus* subsp. *creticus*. *Plant Cell Rep.*, 29(6), 629-641. https://doi.org/10.1007/s00299-010-0849-1

Patterson, B.D., Hatfield, S.G.S., & Knee, M. (1974). Residual effects of controlled atmosphere storage on the production of volatile compounds by two varieties of apples. *J. Sci. Food Agric.*, 25, 843–849. https://doi.org/10.1002/jsfa.2740250714

Pietrysiak, E., & Ganjyal, G.M. (2018). Apple peel morphology and attachment of *Listeria innocua* through aqueous environment as shown by scanning electron microscopy. *Food Control*, 92, 362–369. https://doi.org/10.1016/j.foodcont.2018.04.049

Pizzocaro, F., Torreggiani, D., & Gilardi, G. (1993). Inhibition of apple polyphenoloxidase (PPO) by ascorbic acid, citric acid and sodium chloride. *J. Food Process. Preserv.*, 17(1), 21-30. https://doi.org/10.1111/j.1745-4549.1993.tb00223.x

Poll, L. (1981). Evaluation of 18 apple varieties for their suitability for juice production. *J. Sci. Food Agric.*, 32(11), 1081-1090. https://doi.org/10.1002/jsfa.2740321107

Ponting, J.D., Jackson, R., & Watters, G. (1971). Refrigerated apple slices: effects of pH, sulfites and calcium on texture. *J. Food Sci.*, 36(2), 349-350. https://doi.org/10.1111/j.1365-2621.1971.tb04059.x

Ponting, J.D., Jackson, R., & Watters, G. (1972). Refrigerated apple slices: preservative effects of ascorbic acid, calcium and sulfites. *J. Food Sci.*, 37(3), 434-436. https://doi.org/10.1111/j.1365-2621.1972.tb02657.x

Priepke, P.E.,Wei, L.S., & Nelso, A.I. (1976). Refrigerated storage of prepackaged salad vegetables. *J. Food Sci.*, 41, 379-382. https://doi.org/10.1111/j.1365-2621.1976.tb00624.x

Putnik, P., Roohinejad, S., Greiner, R., Granato, D., Bekhit, A.E.D.A., & Kovačević, D.B. (2017). Prediction and modeling of microbial growth in minimally processed fresh-cut apples packaged in a modified atmosphere: A review. *Food Control*, 80, 411-419. https://doi.org/10.1016/j.foodcont.2017.05.018

Qadri, O.S., Yousuf, B., & Srivastava, A.K. (2015). Fresh-cut fruits and vegetables: Critical factors influencing microbiology and novel approaches to prevent microbial risks—A review. *Cogent Food Agric.*, 1(1), 1121606. https://doi.org/10.1080/23311932.2015.1121606

Qi, H., Hu, W., Jiang, A., Tian, M., & Li, Y. (2011). Extending shelf-life of fresh-cut 'Fuji'apples with chitosan-coatings. *Innov. Food Sci. Emerg. Technol.*, 12(1), 62-66. https://doi.org/10.1016/j.ifset.2010.11.001

Queiroz, C., da Silva, A.J.R., Lopes, M.L.M., Fialho, E., & Valente-Mesquita, V.L. (2011). Polyphenol oxidase activity, phenolic acid composition and browning in cashew apple (*Anacardium occidentale*, L.) after processing. *Food Chem.*, 125(1), 128-132. https://doi.org/10.1016/j.foodchem.2010.08.048

Rabobank International (2010). Retrieved from: http://s3.amazonaws.com/zanranstorage/www.agf.nl/ContentPages/682927373.pdfhttp://s3.amazonaws.com/zanran_storage/www.agf.nl/ContentPages/682927373.pdf

Ragaert, P., Verbeke, W., Devlieghere, F., & Debevere, J. (2004). Consumer perception and choice of minimally processed vegetables and packaged fruits. *Food Qual. Prefer.*, 15(3), 259–270. https://doi.org/10.1016/S0950-3293(03)00066-1

Ragaert, P., Jacxsens, L., Vandekinderen, I., Baert, L., & Devlieghere, F. (2011). Microbiological and safety aspects of fresh-cut fruits and vegetables. In *Advances in fresh-cut fruits and vegetables processing* (pp. 53-75), Martín-Belloso, O., & Soliva-Fortuny, R. (Eds.). CRC Press Taylor & Francis Group: Florida, USA. https://doi.org/10.1201/b10263

Rahman, S.M.E., Jin, Y.G., & Oh, D.H. (2011). Combination treatment of alkaline electrolyzed water and citric acid with mild heat to ensure microbial safety, shelf-life and sensory quality of shredded carrots. *Food Microbiol.,* 28, 484–491. https://doi.org/10.1016/j.fm.2010.10.006

Reineccius, G. (2005). *Flavor chemistry and technology.* CRC press: Boca Raton, Florida, USA. https://doi.org/10.1201/9780203485347

Rizzolo, A., Polesello, A., & Teleky-Vàmossy, G. (1989). CGC/Sensory analysis of volatile compounds developed from ripening apple fruit. *J. High Resolut. Chromatogr.*, 12(12), 824-827. https://doi.org/10.1002/jhrc.1240121214

Rizzolo, A., Grassi, M., & Eccher Zerbini, P. (2006). Influence of harvest date on ripening and volatile compounds in the scab-resistant apple cultivar 'Golden Orange'. *J. Hortic. Sci. Biotechnol.,* 81(4), 681-690. https://doi.org/10.1080/14620316.2006.11512124

Rocculi, P., Romani, S., & Dalla Rosa, M. (2004). Evaluation of physico-chemical parameters of minimally processed apples packed in non-conventional modified atmosphere. *Food Res. Int.,* 37(4), 329-335. https://doi.org/10.1016/j.foodres.2004.01.006

Rocculi, P., Del Nobile, M.A., Romani, S., Baiano, A., & Dalla Rosa, M. (2006). Use of a simple mathematical model to evaluate dipping and MAP effects on aerobic respiration of minimally processed apples. *J. Food Eng.*, 76(3), 334-340. https://doi.org/10.1016/j.jfoodeng.2005.05.034

Rocha, A.M.C.N., Brochado, C.M., & Morais, A.M.M.B. (1998). Influence of chemical treatment on quality of cut apple (cv. Jonagored). *J. Food Qual.*, 21, 13–28. https://doi.org/10.1111/j.1745-4557.1998.tb00500.x

Rojas-Graü M.A., Avena-Bustillos, R.J., Friedman, M., Henika, P.R., Martín-Belloso, O., & McHugh, T.H. (2006). Mechanical, barrier, and antimicrobial properties of apple puree edible films containing plant essential oils. *J. Agric. Food Chem.*, 54(24), 9262–9267. https://doi.org/10.1021/jf061717u

Rojas-Graü, M.A., Grasa-Guillem, R., & Martín-Belloso, O. (2007). Quality changes in fresh-cut Fuji apple as affected by ripeness stage, antibrowning agents, and storage atmosphere. *J. Food Sci.*, 72(1), S036-S043. https://doi.org/10.1111/j.1750-3841.2006.00232.x

Rojas-Graü, M.A., Tapia, M.S., & Martín-Belloso, O. (2008). Using polysaccharide-based edible coatings to maintain quality of fresh-cut Fuji apples. *LWT-Food Sci. Technol*, 41(1), 139-147. https://doi.org/10.1016/j.lwt.2007.01.009

Rowan, D.D., Allen, J.M., Fielder, S., Hunt, M.B. (1997). Deuterium labelling to study aroma biosynthesis in stored apples. In *CA '97 Proceedings Volume 2: apples and pears* (pp. 227-233). Mitcham, E.J. (Ed.). University of California: California, USA.

Rowan, D.D., Allen, J.M., Fielder, S., & Hunt, M.B. (1999). Biosynthesis of straight-chain ester volatiles in Red Delicious and Granny Smith apples using deuterium-labelled precursors. *J. Agric. Food Chem.*, 47(7), 2553-2562. https://doi.org/10.1021/jf9809028

Roy, S., Conway, W.S., Watada, A.E., Sams, C.E., Erbe, E.F., & Wergin, W.P. (1994). Heat treatment affects epicuticular wax structure and postharvest calcium uptake in 'Golden Delicious' apples. *HortSci.*, 29, 1056–1058. https://doi.org/10.21273/HORTSCI.29.9.1056

Rudell, D.R., Mattinson, D.S., Mattheis, J.P., Wyllie, S.G., & Fellman, J.K. (2002). Investigations of aroma volatile biosynthesis under anoxic conditions and in different tissues of 'Redchief Delicious' apple fruit (*Malus domestica* Borkh.). *J. Agric. Food Chem.,* 50(9), 2627-2632. https://doi.org/10.1021/jf011152w

Ruiz-Cruz, S., Alvarez-Parrilla, E., De la Rosa, L.A., Martinez-Gonzalez, A.I., Ornelas-Paz, J.D.J., Mendoza-Wilson, A.M., & Gonzalez-Aguilar, G.A., (2010). Effect of different sanitizers on microbial, sensory and nutritional quality of fresh-cut jalapeno peppers. *Am. J. Agric. Biol. Sci.*, 5, 331-341. https://doi.org/10.3844/ajabssp.2010.331.341

Rux, G., Mahajan, P., Geyer, M., Linke, M., Pant, A., Saengerlaub, S., & Caleb, O. (2015). Application of humidity-regulating tray for packaging of mushrooms. *Postharvest Biol. Technol.*, *108*, 102-110. https://doi.org/10.1016/j.postharvbio.2015.06.010

Rux, G., Caleb, O.J., Fröhling, A., Herppich, W.B., & Mahajan, P.V. (2017). Respiration and storage quality of fresh-cut apple slices immersed in sugar syrup and orange juice. *Food Bioprocess Technol.*, 10 (11), 2081-2091. https://doi.org/10.1007/s11947-017-1980-6

Rux, G., Luca, A., & Mahajan, P.V. (2019a). Changes in volatile organic compounds in the headspace of modified atmosphere packed and unpacked white sausages. *Food Packag. Shelf Life*, 19, 167-173. https://doi.org/10.1016/j.fpsl.2018.12.010

Rux, G., Efe, E., Ulrichs, C., Huyskens-Keil, S., Hassenberg, K., & Herppich, W.B. (2019b). Effects of pre-processing short-term hot-water treatments on quality and shelf life of fresh-cut apple slices. *Foods* 8(12), 653. https://doi.org/10.3390/foods8120653.

Saltveit, M.E. (2019). Respiratory metabolism. In *Postharvest physiology and biochemistry of fruits and vegetables* (pp. 73-91). Yahia, E.M. (Ed.). Woodhead Publishing: Duxford, UK. https://doi.org/10.1016/C2016-0-04653-3

Salvia-Trujillo, L., Rojas-Graü, M.A., Soliva-Fortuny, R., & Martín-Belloso, O. (2015). Use of antimicrobial nanoemulsions as edible coatings: Impact on safety and quality attributes of fresh-cut Fuji apples. *Postharvest Biol. Technol.*, 105, 8-16. https://doi.org/10.1016/j.postharvbio.2015.03.009

Sanz, C., Olias, J.M., & Perez, A.G. (1997). Aroma biochemistry of fruits and vegetables. In *Phytochemistry of fruit and vegetables* (pp. 125-155), Tomas-Barberan, F.A., Robins, R.J. (Eds.). Oxford University Press Inc.: New York, USA.

Sanz, C., & Pérez, A.G. (2010). Plant metabolic pathways and flavor biosynthesis. In *Handbook of fruit and vegetable flavors* (pp. 129-155). Hui, Y.H. (Ed.). John Wiley & Sons, Inc.: Hoboken, New Jersey, USA. https://doi.org/10.1002/9780470622834

Sapers, G.M., & Douglas Jr, F.W. (1987). Measurement of enzymatic browning at cut surfaces and in juice of raw apple and pear fruits. *J. Food Sci.*, 52(5), 1258-1285. https://doi.org/10.1111/j.1365-2621.1987.tb14057.x

Schaffer, R.J., Friel, E.N., Souleyre, E.J., Bolitho, K., Thodey, K., Ledger, S., Bowen, J.H., Ma, J., Nain, B., Cohan, D., Gleave, A.P., Crowhurst, R.N., Janssen, B.J., Yao, J., & Newcomb, R.D. (2007). A genomics approach reveals that aroma production in apple is controlled by ethylene predominantly at the final step in each biosynthetic pathway. *Plant Physiol.*, 144 (4), 1899-1912. https://doi.org/10.1104/pp.106.093765

Schirrmacher, G., & Schempp, H. (2003). Antioxidative potential of flavonoid-rich extracts as new quality marker for different apple varieties. *J. App. Bot. (1995)*, 77(5-6), 163-166.

Schloffer, K., & Linhard, D. (2016). Short time-high temperature hot water shower against Neofabraea rot. In *Ecofruit. 17th International Conference on Organic Fruit-Growing: Proceedings* (pp. 176-179). 15–17 February 2016. FÖKO (Fördergemeinschaft Ökologischer Obstbau eV): Hohenheim, Germany.

Schlüter, O., Foerster, J., Geyer, M., Knorr, D., & Herppich, W.B. (2009). Characterization of high-hydrostatic-pressure effects on fresh produce using chlorophyll fluorescence image analysis. *Food Bioprocess Technol.,* 2(3), 291-299. http://dx.doi/10.1007/s11947-008-0143-1

Schopfer, B., & Brennicke, A. (2010). *Pflanzenphysiologie (7th Ed)*. Springer Spektrum: Unveränd. Nachdruck 2016. http://dx.doi/10.1007/978-3-662-49880-4

Schreier, P. (1984). *Chromatographic studies of biogenesis of plant volatiles*. Alfred Hüthig Verlag GmbH: Heidelberg, Basel, New York.

Schuster, J., & Binder, S. (2005). The mitochondrial branched-chain aminotransferase (AtBCAT-1) is capable to initiate degradation of leucine, isoleucine and valine in almost all tissues in *Arabidopsis thaliana*. *Plant Mol. Biol.*, 57(2), 241-254. https://doi.org/10.1007/s11103-004-7533-1

Schuster, J., Knill, T., Reichelt, M., Gershenzon, J., & Binder, S. (2006). Branched-chain aminotransferase4 is part of the chain elongation pathway in the biosynthesis of methionine-derived glucosinolates in *Arabidopsis*. *Plant Cell*, 18(10), 2664-2679. https://doi.org/10.1105/tpc.105.039339

Schwab, W., Davidovich-Rikanati, R., & Lewinsohn, E. (2008). Biosynthesis of plant-derived flavor compounds. *Plant J.*, 54, 712–732. https://doi.org/10.1111/j.1365-313X.2008.03446.x

Seide, K., Thiem, I., Viedt, H., Wöhlke, A., & Guder, G. (2017). Vorzerkleinertes Obst und Gemüse Beurteilungsgrundlagen, BLL-Leitlinie Untersuchungsergebnisse Niedersächsisches Landesamt für Verbraucherschutz und Lebensmittelsicherheit. Available online: https://www.laves.niedersachsen.de/download/118654/4._Vortrag_LAVES_Lebensmittel-_und_Veterinaerinstitut_Braunschweig_Hannover.pdf (accessed on 17 October 2019).

Shafique, M., Khan, A.S., Malik, A.U., & Shahid, M. (2015). Exogenous application of oxalic acid delays pericarp browning and maintain fruit quality of litchi cv. 'Gola'. *J. Food Biochem.*, 40, 170–179. https://doi.org/10.1111/jfbc.12207

Shao, X.F., Tu, K., Zhao, Y.Z., Chen, L., Chen, Y.Y., & Wang, H. (2007). Effects of prestorage heat treatment on fruit ripening and decay development in different apple cultivars. *J. Hortic. Sci. Biotech.*, 82(2), 297–303. https://doi.org/10.1080/14620316.2007.11512232

Singh, M., & Johnson-Flanagan, A. (1998). Co-ordination of photosynthetic gene expression during low-temperature acclimation and development in *Brassica napus* cv. Jet Neuf leaves. *Plant Sci.*, 135(2), 171-181. https://doi.org/10.1016/S0168-9452(98)00077-6

Siroli, L., Patrignani, F., Serrazanetti, D.I., Tabanelli, G., Montanari, C., Tappi, S., Rocculi, P., Gardini, F., & Lanciotti, R. (2014). Efficacy of natural antimicrobials to prolong the shelf-life of minimally processed apples packaged in modified atmosphere. *Food Control*, 46, 403-411. https://doi.org/10.1016/j.foodcont.2014.05.049

Smit, B.A., Engels, W.J., & Smit, G. (2009). Branched chain aldehydes: production and breakdown pathways and relevance for flavour in foods. *Appl. Microbiol. Biotechnol.*, 81(6), 987-999. https://doi.org/10.1007/s00253-008-1758-x

Smith, A.B., Song, J., & Cameron, A.C. (1998). Modified atmosphere packaged cut iceberg: effect of temperature and O_2 partial pressure on respiration and quality. *J. Agric. Food Chem.*, 46, 4556–4562. https://doi.org/10.1021/jf980208s

Soliva-Fortuny, R.C., Grigelmo-Miguel, N., Odriozola-Serrano, I., Gorinstein, S., & Martín-Belloso, O. (2001). Browning evaluation of ready-to-eat apples as affected by modified atmosphere packaging. *J. Agric. Food Chem.*, 49(8), 3685-3690. https://doi.org/10.1021/jf010190c

Soliva-Fortuny, R.C., Oms-Oliu, G., & Martín-Belloso, O. (2002). Effects of ripeness stages on the storage atmosphere, color, and textural properties of minimally processed apple slices. *J. Food Sci.*, 67(5), 1958-1963. https://doi.org/10.1111/j.1365-2621.2002.tb08752.x

Soliva-Fortuny, R.C., & Martín-Belloso, O. (2003a). New advances in extending the shelf-life of fresh-cut fruits: a review. *Trends Food Sci. Technol.*, 14, 341-353. https://doi.org/10.1016/S0924-2244(03)00054-2

Soliva-Fortuny, R.C., Lluch, M.A., Quiles, A., Grigelmo-Miguel, N., & Martín-Belloso, O. (2003b). Evaluation of textural properties and microstructure during storage of minimally processed apples. *J. Food Sc.*, 68(1), 312-317. https://doi.org/10.1111/j.1365-2621.2003.tb14158.x

Soliva-Fortuny, R.C., Elez-Martínez, P., & Martín-Belloso, O. (2004). Microbiological and biochemical stability of fresh-cut apples preserved by modified atmosphere packaging. *Innov. Food Sci. Emerg.*, 5(2), 215-224. https://doi.org/10.1016/j.ifset.2003.11.004

Soliva-Fortuny, R.C., Ricart-Coll, M., & Martín-Belloso, O. (2005). Sensory quality and internal atmosphere of fresh-cut Golden Delicious apples. *Int. J. Food Sci. Tech.,* 40(4), 369-375. https://doi.org/10.1111/j.1365-2621.2004.00934.x

Song, J., & Bangerth, F. (1996). The effect of harvest date on aroma compound production from 'Golden Deli cious' apple fruit and relationship to respiration and ethylene production. *Postharvest Biol. Technol.,* 8, 259-269. https://doi.org/10.1016/0925-5214(96)00020-8

Song, J., Fan, L., Forney, C.F., & Jordan, M.A. (2001). Using volatile emissions and chlorophyll fluorescence as indicators of heat injury in apples. *J. Am. Soc. Hortic. Sci.*, 126(6), 771-777. https://doi.org/10.21273/JASHS.126.6.771

Song, J., & Bangerth, F. (2003). Fatty acids as precursors for aroma volatile biosynthesis in pre-climacteric and climacteric apple fruit. *Postharvest Biol. Technol.,* 30(2), 113-121. https://doi.org/10.1016/S0925-5214(03)00098-X

Spadoni, A., Guidarelli, M., Phillips, J., Mari, M., & Wisniewski, M. (2015). Transcriptional profiling of apple fruit in response to heat treatment: Involvement of a defense response during *Penicillium expansum* infection. *Postharvest Biol. Technol.*, 101, 37–48. https://doi.org/10.1016/j.postharvbio.2014.10.009

Tapia de Daza, M.S., Alzamora, S.M., Chanes, J.W., & Gould, G. (1996). Combination of preservation factors applied to minimal processing of foods. *Crit. Rev. Food Sci. Nutr.*, 36, 629–659. https://doi.org/10.1080/10408399609527742

Tappi, S., Berardinelli, A., Ragni, L., Dalla Rosa, M., Guarnieri, A., & Rocculi, P. (2014). Atmospheric gas plasma treatment of fresh-cut apples. *Innov. Food Sci. Emerg. Technol.*, 21, 114-122. https://doi.org/10.1016/j.ifset.2013.09.012

Teixido, N., Usall, J., & Vinas, I. (1999). Efficacy of preharvest and postharvest *Candida sake* biocontrol treatments to prevent blue mould on apples during cold storage. *Int. J. Food Microbiol.*, 50(3), 203-210. https://doi.org/10.1016/S0168-1605(99)00105-1

Toivonen, P.M., & DeEll, J.R. (2002). Physiology of fresh-cut fruits and vegetables. In *Fresh-cut fruits and vegetables: Science, technology, and market* (pp. 91-123). Lamikanra, O. (Ed.). CRC Press, Boca Raton, FL, USA. https://doi.org/10.1201/9781420031874

Toivonen, P.M. (2004). Postharvest storage procedures and oxidative stress. *HortSci.*, 39(5), 938-942.

Toivonen, P.M., & Brummell, D.A. (2008). Biochemical bases of appearance and texture changes in fresh-cut fruit and vegetables. *Postharvest Biol. Technol.*, *48*(1), 1-14. https://doi.org/10.1016/j.postharvbio.2007.09.004

Tortoe, C., Orchard, J., & Beezer, A. (2007). Prevention of enzymatic browning of apple cylinders using different solutions. *Int. J. Food Sci. Technol.*, 42(12), 1475-1481. https://doi.org/10.1111/j.1365-2621.2006.01367.x

Tournas, V.H., Heeres, J., & Burgess, L., (2006). Moulds and yeasts in fruit salads and fruit juices. *Food Microbiol.*, 23, 684-688. https://doi.org/10.1016/j.fm.2006.01.003

Tressl, R., & Drawet, F. (1973). Biogenesis of banana volatiles. *J. Agric. Food Chem.*, 21, 560-565. https://doi.org/10.1021/jf60188a031

Trierweiler, B., Schirmer, H., & Tauscher, B. (2003). Hot water treatment to control *Gloeosporium* disease on apples during long-term storage. *J. Appl. Bot.*, 77, 156–159.

Tzin, V., & Galili, G. (2010). New insights into the shikimate and aromatic amino acids biosynthesis pathways in plants. *Mol. Plant*, 3(6), 956-972. https://doi.org/10.1093/mp/ssq048

Uyttendaele, M., Neyts, K., Vanderswalmen, H., Notebaert, E., & Debevere, J. (2004). Control of *Aeromonas* on minimally processed vegetables by decontamination with lactic acid, chlorinated water, or thyme essential oil solution. *Int. J. Food Microbiol.*, 90(3), 263-271. https://doi.org/10.1016/S0168-1605(03)00309-X

Van Buggenhout, S., Sila, D.N., Duvetter, T., Van Loey, A., & Hendrickx, M. (2009). Pectins in processed fruits and vegetables: Part III—Texture engineering. *Compr. Rev. Food Sci. Food Saf.*, 8(2), 105-117. https://doi.org/10.1111/j.1541-4337.2009.00072.x

Van den Dool, H., & Kratz, P.D. (1963). A generalization of the retention index system including linear temperature programmed gas-liquid partition chromatography. *J. Chromatogr. A,* 11, 463-471. https://doi.org/10.1016/s0021-9673(01)80947-x

Van Rijswick, C. (2010). EU Fresh-cut Fruits and Vegetables Market Update. *Rabobank Industry Note 246*. Retrieved from: https://docplayer.net/21371633-Savid-the-eu-fresh-cut-fruits-and-vegetables-market.html

Varoquaux, P., Lecendre, I., Varoquaux, F., & Souty, M. (1990). Change in firmness of kiwifruit after slicing. *Sciences des Aliments*, 10(1), 127-139.

Varoquaux, P., & Wiley, R.C. (1994). Biological and biochemical changes in minimally processed refrigerated fruits and vegetables. In *Minimally processed refrigerated fruits & vegetables* (pp. 226-268). Springer: Boston, MA. https://doi.org/10.1007/978-1-4615-2393-2_6

Verlinden, S., & Garcia, J.J.V. (2004). Sucrose loading decreases ethylene responsiveness in carnation (*Dianthus caryophyllus* cv. White Sim) petals. *Postharvest Biol. Technol.*, 31(3), 305-312. https://doi.org/10.1016/j.postharvbio.2003.09.010

Vincent, J.F.V. (1994). Texture of plants. In *Vegetables and vegetable products* (pp. 57-72). Linskens, H.F., Jackson, J.F. (Eds.). Springer: Berlin, Heidelberg, Germany. https://doi.org/10.1007/978-3-642-84830-8

von Willert, D.J., Matyssek, R., & Herppich, W.B. (1995). *Experimentelle Pflanzenökologie: Grundlagen und Anwendungen*. Georg Thieme Verlag, Stuttgart. ISBN 3-13-134401-6. https://doi.org/10.1007/978-3-662-53465-6

Watada, A.E., & Qi, L. (1999). Quality of fresh-cut produce. *Postharvest Biol. Technol.*, 15(3), 201-205. https://doi.org/10.1016/S0925-5214(98)00085-4

Watkins, C.B. (2000). Responses of horticultural commodities to high carbon dioxide as related to modified atmosphere packaging. *Hort-Technol.,* 10, 501–506. https://doi.org/10.21273/HORTTECH.10.3.501

Weemaes, C.A., Ludikhuyze, L.R., Van den Broeck, I., Hendrickx, M.E., & Tobback, P.P. (1998). Activity, electrophoretic characteristics and heat inactivation of polyphenoloxidases from apples, avocados, grapes, pears and plums. *LWT-Food Sci. Technol.*, 31(1), 44-49. https://doi.org/10.1006/fstl.1997.0302

Whitaker, J.R. (1994). *Principles of Enzymology for the Food Sciences (2nd Ed.)*. Routledge: Boca Raton, FL, USA. https://doi.org/10.1201/9780203742136

Wills, R.B.H., McGlasson, W.B. (1971). Effect of stor age temperature on apple volatiles associated with low temperature breakdown. *J. Hortic. Sci.*, 46, 115-120. https://doi.org/10.1080/00221589.1971.11514390

Woodstock, L.W., & Taylorson, R.B. (1981). Ethanol and acetaldehyde in imbibing soybean seeds in relation to deterioration. *Plant Physiol.*, 67(3), 424-428. https://doi.org/10.1104/pp.67.3.424

Wooltorton, L.S.C., Jones, J.D., Hulme, A.C. (1965). Genesis of ethylene in apples. *Nature* 207, 999. https://doi.org/10.1038/207999b0

World Health Organization. (1998). Food safety issues: Surface decontamination of fruits and vegetables eaten raw: a review. *WHO/FSF/FOS/98.2*. World Health Organization: Geneva. https://apps.who.int/iris/handle/10665/64435

Wright, A.H., Delong, J.M., Arul, J., & Prange, R.K. (2015). The trend toward lower oxygen levels during apple (*Malus* x *domestica* Borkh) storage – A review. *J. Hortic. Sci. Biotech.*, 90, 1–13. https://doi.org/10.1080/14620316.2015.11513146

Yabumoto, K., & Jennings, W.G. (1977). Volatile constituents of cantaloupe, *Cucumis melo*, and their biogenesis. *J. Food Sci.*, 42(1), 32-37. https://doi.org/10.1111/j.1365-2621.1977.tb01212.x

Yahia, E.M. (2006). Controlled atmosphere, modified atmosphere and modified atmosphere packaging for vegetables. *Stewart Postharvest Rev.*, 2(5–6), 1–6. https://doi.org/10.2212/spr.2006.5.5

Yang, S.F. (1981). Biosynthesis of ethylene and its regulation. In *Recent advances in the biochemistry of fruit and vegetables* (pp. 89-106). J. Friend, & M. J. C. Rhodes (Eds.). Academic Press: London, UK.

Yearsley, C.W., Banks, N.H., Ganesh, S., & Cleland, D.J. (1996). Determination of lower oxygen limits for apple fruit. *Postharvest Biol. Technol.*, 8(2), 95-109. https://doi.org/10.1016/0925-5214(96)00064-6

Young, H., Gilbert, J.M., Murray, S.H., & Ball, R.D. (1996). Causal effects of aroma compounds on Royal Gala apple flavours. *J. Sci. Food Agric.*, 71(3), 329-336. https://doi.org/10.1002/(SICI)1097-0010(199607)71:3<329::AID-JSFA588>3.0.CO;2-8

Zepka, L.Q., Garruti, D S., Sampaio, K.L., Mercadante, A.Z., & Da Silva, M.A.A. (2014). Aroma compounds derived from the thermal degradation of carotenoids in a cashew apple juice model. *Food Res. Int.*, 56, 108-114. https://doi.org/10.1016/j.foodres.2013.12.015

Zhu, Y.I., Pan, Z., & McHugh, T.H. (2007). Effect of dipping treatments on color stabilization and texture of apple cubes for infrared dry-blanching process. *J. Food Process. Pres.*, 31(5), 632-648. https://doi.org/10.1111/j.1745-4549.2007.00154.x

Zhuang, H., Barth, M.M., & Hankinson, T.R. (2003). Microbial safety, quality, and sensory aspects of fresh-cut fruits and vegetables. In *Microbial Safety of Minimally Processed Foods* (pp. 255–278). Novak, J.S., Sapers, G.M., & Juneja V.K. (Eds.). CRC Press: Boca Raton, Fla, USA. https://doi.org/10.1201/9781420031850

Danksagung

An dieser Stelle möchte ich allen Personen meinen großen Dank aussprechen, die mich bei der Bearbeitung meines Promotionsthemas und der Anfertigung meiner Dissertationsschrift unterstützt haben. Ich bedaure es sehr, hier nicht alle namentlich erwähnen zu können.

Ich danke der Humboldt-Universität zu Berlin für die mir gegebene Möglichkeit, eine Promotion durchzuführen. Prof. Dr. Dr. Christian Ulrichs danke ich hierbei vor allem für seinen wesentlichen Beitrag bei der Ausarbeitung meiner Forschungsfrage, für die freundlichen und konstruktiven Besprechungen, für die gewährten Freiräume bei der Anfertigung der Arbeit, aber auch für das mir entgegengebrachte Vertrauen. Erheblichen Dank möchte ich Prof. Dr. Susanne Huyskens-Keil für die zusätzliche und entgegenkommende Betreuung, die vielen nützlichen Anregungen sowie lieben, aufmunternden Worten und der umfassenden und zeitintensiven Durchsicht meiner Arbeit aussprechen.

Mein besonderer Dank gilt Dr. Werner Herppich für die hervorragende Betreuung, die vielfältige Unterstützung bei der Bearbeitung des Themas und seiner Zeit für ausführliche gemeinsame Diskussionen. Ohne seinen Zuspruch, seine Geduld und Zuversicht wäre diese Dissertation nie fertiggestellt worden. Auf meinem Weg hat er mich mit zahllosen produktiven sowie freundschaftlichen Gesprächen begleitet und so meine fachliche, aber auch persönliche Entwicklung wesentlich geprägt. Ich hätte mir keinen besseren Kollegen wünschen können.

Ich bedanke mich auch bei Dr. Pramod Mahajan, der mich motivierte, die Promotion überhaupt erst in Angriff zu nehmen, entscheidend bei der Wahl des Themas mitwirkte und mir durch viele Freiräume ermöglichte neben meiner eigentlichen Projektarbeit meine Dissertation anzufertigen. Des Weiteren möchte ich mich bei Dr. Karin Hassenberg bedanken, die meiner Promotion eine neue Perspektive gab, diese mit Rat und Tat begleitete sowie deren strukturierte Arbeitsweise und fröhliche Art mich stets inspirierte. Nicht zuletzt möchte ich mich auch bei Kathrin Ilte bedanken, mit der ich mich in einer sehr engen Zusammenarbeit stets sowohl fachlich zur VOC-Analytik als auch freundschaftlich austauschen konnte und die immer ein offenes Ohr für mich hatte. Bei der tatkräftigen Unterstützung meiner praktischen Arbeiten sollen einige Personen nicht ungenannt bleiben und verdienen ebenfalls meinen Dank: David Sakowsky für zahllose Hilfestellungen beim Auf- und Umbau meiner Versuchsanordnungen, Gabriele Wegner und Corinna Rolleczek für ihren Beitrag bei den chemischen und physikalischen Analysen, Janett Schiffmann, Imke Handke und Susanne Klocke für die Durchführung der mikrobiologischen Untersuchungen, Christian Regen, Ingo Truppel und Stefan Elwart für ihre technische Unterstützung.

Dem Leibniz-Institut für Agrartechnik und Bioökonomie, an dem ich den größten Teil meiner Arbeit angefertigt habe, danke ich für die Bereitstellung der nötigen Infrastruktur und der finanzielle Unterstützung durch ein Promotionsstipendium, durch das ich mich in einer entscheiden Phase auf die Fertigstellung meiner Arbeit konzentrieren konnte. Ich danke allen Mitarbeitenden der Abteilung Technik im Gartenbau des ATBs für die angenehme Arbeitsatmosphäre, den fachlichen Austausch sowie vielen prägenden Erfahrungen während der vergangen Jahre. Bei dieser Gelegenheit möchte ich zudem Dr. Martin Geyer hervorheben und meinen herzlichen Dank äußern. Er hatte meine Arbeit ebenfalls erst ermöglicht, durch seine Unterstützung gefördert und mich mit starkem Entgegenkommen während der gesamten Promotion entlastet.

Einen späten Dank will ich an dieser Stelle auch der Beuth-Hochschule und ihren Mitarbeitenden aussprechen, die mir durch ihre Arbeit meine akademischen Grundlagen vermittelten und somit diesen Lebensweg ermöglicht haben.

Liebevoll bedanke ich mich bei meiner Partnerin Charlotte für ihre neugierigen Fragen, konstruktiven Anmerkungen und die emotionale Unterstützung und Motivation, vor allem in den belastenden Phasen meiner Promotion.

Ich wünsche mir, dass nach mir noch viele andere die gleiche Unterstützung durch so freundliche und kompetente Menschen erhalten können, wie ich ihnen begegnen durfte. Weiterhin hoffe ich, dass auch zukünftig Möglichkeiten der finanziellen Unterstützung, wie allen voran durch die Förderung nach dem BaFöG aber auch durch Stipendien existieren, die oft die Grundlage bilden, überhaupt eine akademische Laufbahn einschlagen zu können.

Declaration – Eidesstattliche Erklärung

Erklärung:

Hiermit erkläre ich, die Dissertation selbstständig und nur unter Verwendung der angegebenen Hilfen und Hilfsmittel angefertigt zu haben.

Ich habe mich anderwärts nicht um einen Doktorgrad beworben und besitze keinen entsprechenden Doktorgrad.

Ich erkläre, dass ich die Dissertation oder Teile davon nicht bereits bei einer anderen wissenschaftlichen Einrichtung eingereicht habe und dass sie dort weder angenommen noch abgelehnt wurde.

Ich erkläre die Kenntnisnahme der dem Verfahren zugrundeliegenden Promotionsordnung der Lebenswis-senschaftlichen Fakultät der Humboldt-Universität zu Berlin vom 5. März 2015.

Weiterhin erkläre ich, dass keine Zusammenarbeit mit gewerblichen Promotionsbearbeiterinnen/Promoti-onsberatern stattgefunden hat und dass die Grundsätze der Humboldt-Universität zu Berlin zur Sicherung guter wissenschaftlicher Praxis eingehalten wurden.

Declaration:

I hereby declare that I completed the doctoral thesis independently based on the stated resources and aids.

I have not applied for a doctoral degree elsewhere and do not have a corresponding doctoral degree.

I have not submitted the doctoral thesis, or parts of it, to another academic institution and the thesis has not been accepted or rejected.

I declare that I have acknowledged the Doctoral Degree Regulations, which underlie the procedure of the Faculty of Life Sciences of Humboldt-Universität zu Berlin, as amended on 5th March 2015.

Furthermore, I declare that no collaboration with commercial doctoral degree supervisors took place, and that the principles of Humboldt-Universität zu Berlin for ensuring good academic practice were abided by.

Berlin, den 29.06.2020 Guido Rux

www.ingramcontent.com/pod-product-compliance
Ingram Content Group UK Ltd.
Pitfield, Milton Keynes, MK11 3LW, UK
UKHW022000190726
13853UKWH00004B/1653

9 783736 975644